Florian Dürr

Laser-induced stress changes in optical fibers

Florian Dürr

Laser-induced stress changes in optical fibers

About the acquisition and interpretation of in-fiber stress changes caused by laser-irradiation

Südwestdeutscher Verlag für Hochschulschriften

Impressum/Imprint (nur für Deutschland/ only for Germany)
Bibliografische Information der Deutschen Nationalbibliothek: Die Deutsche Nationalbibliothek verzeichnet diese Publikation in der Deutschen Nationalbibliografie; detaillierte bibliografische Daten sind im Internet über http://dnb.d-nb.de abrufbar.
Alle in diesem Buch genannten Marken und Produktnamen unterliegen warenzeichen-, marken- oder patentrechtlichem Schutz bzw. sind Warenzeichen oder eingetragene Warenzeichen der jeweiligen Inhaber. Die Wiedergabe von Marken, Produktnamen, Gebrauchsnamen, Handelsnamen, Warenbezeichnungen u.s.w. in diesem Werk berechtigt auch ohne besondere Kennzeichnung nicht zu der Annahme, dass solche Namen im Sinne der Warenzeichen- und Markenschutzgesetzgebung als frei zu betrachten wären und daher von jedermann benutzt werden dürften.

Verlag: Südwestdeutscher Verlag für Hochschulschriften Aktiengesellschaft & Co. KG
Dudweiler Landstr. 99, 66123 Saarbrücken, Deutschland
Telefon +49 681 37 20 271-1, Telefax +49 681 37 20 271-0, Email: info@svh-verlag.de
Zugl.: Lausanne, EPFL, Dissertation, 2005

Herstellung in Deutschland:
Schaltungsdienst Lange o.H.G., Berlin
Books on Demand GmbH, Norderstedt
Reha GmbH, Saarbrücken
Amazon Distribution GmbH, Leipzig
ISBN: 978-3-8381-0477-5

Imprint (only for USA, GB)
Bibliographic information published by the Deutsche Nationalbibliothek: The Deutsche Nationalbibliothek lists this publication in the Deutsche Nationalbibliografie; detailed bibliographic data are available in the Internet at http://dnb.d-nb.de.
Any brand names and product names mentioned in this book are subject to trademark, brand or patent protection and are trademarks or registered trademarks of their respective holders. The use of brand names, product names, common names, trade names, product descriptions etc. even without a particular marking in this works is in no way to be construed to mean that such names may be regarded as unrestricted in respect of trademark and brand protection legislation and could thus be used by anyone.

Publisher:
Südwestdeutscher Verlag für Hochschulschriften Aktiengesellschaft & Co. KG
Dudweiler Landstr. 99, 66123 Saarbrücken, Germany
Phone +49 681 37 20 271-1, Fax +49 681 37 20 271-0, Email: info@svh-verlag.de

Copyright © 2009 by the author and Südwestdeutscher Verlag für Hochschulschriften Aktiengesellschaft & Co. KG and licensors
All rights reserved. Saarbrücken 2009

Printed in the U.S.A.
Printed in the U.K. by (see last page)
ISBN: 978-3-8381-0477-5

Abstract

The precise determination of residual stress profiles in optical fibers allows drawing conclusions about radiation-induced density-changes of the fibers doped core glass. The corresponding index changes partially contribute to the phenomenon of photosensitivity, i.e. the modification of the refractive index of a glass through the irradiation with an appropriate laser source. Photosensitivity is widely exploited for the fabrication of fiber Bragg gratings which are in-fiber filter devices whose spectral response can be designed with great versatility. Fiber Bragg gratings find broad application nowadays in the field of telecommunication as well as in sensing applications.

The polarimetric assessment of the integrated transverse fiber birefringence serves as a means to gather information about both stress and strain profiles of the fiber. Birefringence is introduced in the optical fiber due to the photo-elastic effect. In this work, a highly sensitive polariscope with submicron resolution has been developed. The captured birefringence data has been interpreted to be basically strain and not stress-induced, which allowed the proper discrimination between an inelastic, stress-free strain contribution and a strain part essentially accompanied by elastic stress. In earlier interpretations of the polarimetric data, all birefringence had been attributed to be purely stress-induced, which excluded the existence of inelastic strain and even led to contradictory results in some cases. The reinterpretation in terms of strain, however, results in a consistent picture of both stress and strain-induced birefringence.

The polariscope has been used in the following to determine radiation-induced stress changes for different optical fibers and irradiation sources. Femtosecond-laser induced core stress changes in SMF-28® standard telecommunication fibers have been measured tomographically and correlated with corresponding index changes. Furthermore, stress has been correlated to the refractive index change in highly germanium-doped fibers irradiated with cw-irradiation at 244 nm. The annealing behavior of stress, strain and index in this fiber has been studied. Approximately the same linear correlation between stress and index change was found for all germanium-doped fibers under investigation. In addition, stress changes were measured for phosphorus- and nitrogen-doped fibers drawn at different tensions before and after irradiation with an ArF excimer-laser. The impact of the drawing force on fiber stress and inelastic strain anisotropy has been determined as well as radiation-induced stress modifications. The observed stress increase has been compared to those obtained for germanium-doped fibers.

Atomic Force Microscopy of cleaved and subsequently etched fiber endfaces is a powerful means to gather topographic information on a nanometer scale due to differential etching of the doped core and the undoped silica cladding. Within this work, the technique has been used to acquire information about etch rate changes in fibers drawn from the same nitrogen-doped preform with different drawing conditions. Furthermore, the impact of UV-irradiation on the etch rate has been determined for phosphorus- as well as nitrogen-doped fibers. Comparison of the results with stress measurements allowed the attribution of the etch rate changes to a modification of the color center population.

Contents

1 Introduction   **9**
1.1 State of the art 9
    1.1.1 Photo-induced stress and index changes in optical fibers   9
    1.1.2 Stress measurements in optical fibers 10
1.2 Motivation and thesis outline 11

2 Residual stresses and strains in optical fibers   **17**
2.1 Elastic stresses in optical fibers 18
    2.1.1 Plane strain analysis 18
    2.1.2 Stresses in cylinders with radial symmetry 20
    2.1.3 Stresses in cylinders of arbitrary cross section 22
    2.1.4 Thermal stresses 24
    2.1.5 UV-induced stresses 25
2.2 Drawing-induced stresses and strains: Viscoelasticity 28
    2.2.1 Drawing-induced stresses 28
    2.2.2 Inelastic strains in optical fibers 30
2.3 Conclusions 33

3 Photosensitivity of optical waveguides   **39**
3.1 Origins of photosensitivity 40
    3.1.1 Color-center model 41
    3.1.2 Compaction model 41
3.2 Photoelasticity and Photoplasticity 43
    3.2.1 Historical overview 43
    3.2.2 Stress- and strain-induced refractive index changes . . 44
3.3 Irradiation-induced index changes in optical fibers 46
    3.3.1 Densification-induced refractive index changes 46
    3.3.2 Elastic response and net index change 48
3.4 Measurement of photosensitivity 50
    3.4.1 Bragg grating inscription 50
    3.4.2 Atomic Force Microscopy 53
3.5 Summary and Conclusions 54

4 Stress and strain profile determination in optical fibers 63
- 4.1 Review of existing measurement techniques 63
- 4.2 Theory 66
  - 4.2.1 Integrated phase retardation 66
  - 4.2.2 Axial symmetric stress profiles: Abel-transform 68
  - 4.2.3 Asymmetric stress profiles: optical tomography 71
- 4.3 Stress measurement setup 73
  - 4.3.1 Setup description 73
  - 4.3.2 Setup calibration 79
  - 4.3.3 Accuracy and Repeatability 81
  - 4.3.4 Spatial resolution 83
- 4.4 Conclusions 84

5 Laser-induced stress changes in optical fibers 91
- 5.1 Germanium-doped fibers 92
  - 5.1.1 Femtosecond-laser induced stress changes 92
  - 5.1.2 Annealing-induced stress changes in UV-irradiated germanium-doped fibers 99
- 5.2 Phosphorus- and Nitrogen-doped fibers 108
  - 5.2.1 Fibers under investigation 108
  - 5.2.2 Stress and inelastic strain changes with drawing tension 109
  - 5.2.3 UV-induced stress changes 114
- 5.3 Summary of results and discussion 118

6 Fiber characterization using Atomic Force Microscopy 127
- 6.1 Phosphorus-doped fibers 127
  - 6.1.1 Experiment 127
  - 6.1.2 Etch dynamics of the pristine fiber 129
  - 6.1.3 Hydrogen- and UV-induced etch rate changes 130
  - 6.1.4 Discussion of results 131
  - 6.1.5 Conclusions 134
- 6.2 Nitrogen-doped fibers 135
  - 6.2.1 Experiment 135
  - 6.2.2 Etch dynamics of the pristine fiber 136
  - 6.2.3 UV-induced etch rate changes 136
  - 6.2.4 Discussion 137
  - 6.2.5 Conclusions 141
- 6.3 Summary and Discussion 141

7 Conclusions 147
- 7.1 Summary 147
- 7.2 Future work 148

Chapter 1

Introduction

1.1 State of the art

1.1.1 Photo-induced stress and index changes in optical fibers

A permanent change in refractive index with a corresponding modification of the absorption spectrum induced by light irradiation into transparent media is referred to as photosensitivity [1]. The potentiality to use this effect to photo-inscribe grating structures in germanosilicate fibers has been reported for the first time by Hill *et al.* in 1978 [2]. Since the introduction of the side-writing interference technique by Meltz *et al.* in 1989 [3], in-fiber Bragg gratings find increasing application in wavelength-division multiplexed fiber optic communications [4] as well as in optical sensor systems [5], as it decouples the grating period from the wavelength of the illuminating coherent laser beam.

Although the discovery of photosensitivity in optical fibers by Hill is already more than 20 years ago, there is still no complete physical explanation for it at present. The irradiation of doped silica is always accompanied by changes in its UV and VUV absorption spectrum, which are linked to changes in refractive index by the Kramers-Kronig relationship. The changes in the UV can be related to the modification of defects introduced in the glass during the fiber fabrication process [6], whereas changes in the VUV are introduced by a densification of the glass [7]. As the changes in the UV were easier to identify than those in the VUV, it was first concluded that only the defect modification accounts for the phenomenon of photosensitivity [8–10]. However, this so-called color-center model was found to fail for a number of experiments [11, 12], so it became evident that also the effect of densification has to be considered.

Bernardin *et al.* were the first to suggest a structural alteration in the mechanical nature of the glass after irradiation as a second reason to cause photosensitivity [13]. They proposed the collapse of higher order ring structures to result in a densification of the doped core glass. Densification is linearly correlated with an index increase. Experimental evidence for this so-called compaction model was found by scanning the surface of irradiated preform samples with atomic force microscopy [14] as well as interferometric profilometry [15].

The densification of the core glass is in general accompanied by a modification of the residual stress profile introduced into the fiber during its fabrication process. A modification of stresses again leads to a change in refractive index due to the photo-elastic effect. Wong *et al.* suggested the relief of tensile core stress as main reason for the index increase in germanosilicate fibers [16]. However, Fonjallaz *et al.* refuted the stress relief model by observing an increase in core stress independent of the initial stress state (tensile or compressive) for germanosilicate fibers after UV-irradiation [17]. As an increase in core stress corresponds to a decrease in refractive index, it was concluded that the densification and color-center induced index change must exceed the photo-elastic one to obtain an overall positive index [18].

According to the current understanding of photosensitivity, both color-center changes and densification contribute to the overall index change. The amount of each contribution might vary strongly as a function of fiber dopant and pre-treatment as well as with different irradiating sources.

1.1.2 Stress measurements in optical fibers

Both stresses and inelastic strains introduce birefringence into optical fibers [19]. Historically, however, the measured birefringence has been attributed to be solely stress-induced. Only recently, Yablon *et al.* reported about the appearance of inelastic strain in optical fibers [20], which will also impact the birefringence data. The review about existing measurement techniques presented here thus only mentions stress-induced birefringence. Within this thesis, however, we present an interpretation of birefringence data including both the contribution due to inelastic strain and stress.

For cylindrical symmetries, the axial component of the stress tensor can be evaluated from the accumulated transverse birefringence by an Abel-inversion [21]. The classical polarimetric method to quantify birefringence is to put the sample between two linear polarizers with orthogonal optical axes. This method can in principal also be applied to optical fibers [22]. However, the method fails if only a small amount of birefringence ($\sim 10^{-2}$ rad) is introduced by the sample, as for residual stresses in optical fibers. Therefore, in 1982, Chu and Whitbread introduced a measurement setup [23] based on the Sénarmont compensation method [24]. Chu's method for stress measurements in optical fibers has been improved and further developed by several researchers. Abe *et al.* performed tomographic measurements of axial stress by acquiring retardation profiles for different rotation angles of the fiber [25]. The stress is deduced from the retardation profile by an inverse Radon-transform [26]. Raine proposed a half-shade principle [27] to enhance the sensitivity of the axial stress measurement [28]. He also showed the capability of tomographic measurements for his setup [29]. Park *et al.* introduced a setup with a fixed analyzer and a rotating polarizer, which is reciprocal to Chu's setup [30]. This makes it possible to locate the fixed polarizer directly behind the optical fiber and thus reduce wavefront errors. Additionally, an analog CCD camera is used instead of a photodiode for data acquisition. The setup has also been used to determine the stress tensor of a polarization-maintaining fiber by photo-elastic tomography [31].

1.2 Motivation and thesis outline

The primary goal of this thesis work is the measurement as well as the interpretation of stresses and stress related index changes in optical fibers caused by the interaction with laser-irradiation. By comparison of the densification induced index change with the overall index change, different contributions to the observed index modification can be identified, improving the understanding of fundamental laser-induced processes. The thesis thus consists of an engineering part including the development of measuring techniques to acquire the desired data, and an analyzing part related to materials science.

Chapter 2 summarizes the origins of stresses and inelastic strains in optical fibers. Analytical solutions for their quantification in circular symmetric geometries are presented as well as guidelines for their calculations in asymmetric structures. All components of the stress and strain tensor can essentially be derived only from their axial components.

Chapter 3 summarizes the different contributions to irradiation-induced refractive index changes in optical fibers. Particularly, index modifications due to a compaction of the core glass are discussed. Therefore, the photo-elastic effect is described as well as its extension to the plastic regime. The properties of fiber Bragg gratings are resumed to show their potential to quantify radiation-induced refractive index changes. Finally, a short review about the determination of fiber properties by scanning etched fiber end-faces with an atomic force microscope is given.

Chapter 4 focuses on the development and characterization of the measurement setup. The setup has been slightly modified in comparison to Park's setup [30] to get a higher freedom for the alignment of the fiber. The retardation data obtained is interpreted both in terms of stress and inelastic strain.

Chapter 5 presents the measurement results obtained for different fibers and irradiation conditions. For germanium-doped fibers, stress changes due to irradiation with both femtosecond-laser pulses and continuous wave UV-irradiation are presented. The annealing behavior of UV-induced stress changes is described. For phosphorus- and nitrogen-doped fibers, the impact of the drawing tension on stress as well as inelastic strain is demonstrated. Finally, the radiation-induced densification is compared for the three different fiber dopants.

Chapter 6 summarizes the results for etch rate differences in optical fiber cores due to fiber processing conditions and UV-irradiation. The etch rates were determined using an atomic force microscope. The observed changes in etch rate are attributed to corresponding modifications in the color-center population.

Chapter 7 resumes the most important results of the thesis and gives a short motivation for future work.

Bibliography

[1] Raman Kashyap. *Fiber Bragg Gratings*. Academic Press, 1999.

[2] K. O. Hill, Y. Fujii, D. C. Johnson, and B. S. Kawasaki. Photosensitivity in optical fiber waveguides: Application to reflection filter fabrication. *Applied Physics*, 32(10):647–649, 1978.

[3] G. Meltz, W. W. Morey, and W. H. Glenn. Formation of Bragg gratings in optical fibers by a transverse holographic method. *Optics Letters*, 14(15):823–825, 1989.

[4] V. Mizrahi and T. Erdogan. UV-written fiber gratings for optical communications systems. In *Proc. European Conf. on Optical Communication, ECOC'94*, 1994. 991-6 vol.2.

[5] A. D. Kersey, M. A. Davis, H. J. Patrick, M. LeBlanc, K. P. Koo, C. G. Askins, M. A. Putnam, and E. J. Friebele. Fiber grating sensors. *Journal of Lightwave Technology*, 15(8):1442–1463, 1997.

[6] V. B. Neustruev. Colour centres in germanosilicate glass and optical fibres. *Journal of Physics: Condensed Matter*, 6:6901–6936, 1994.

[7] C. Z. Tan and J. Arndt. The refractive index of silica glass and its dependence on pressure, temperature, and the wavelength of the incident light. In H. S. Nalwa, editor, *Silicon based materials and devices*. Academic Press, 2001.

[8] D. L. Williams, S. T. Davey, R. Kashyap, J. R. Armitage, and B. J. Ainslie. Direct observation of UV induced bleaching of 240 nm absorption band in photosensitive germanosilicate glass fibres. *Electronics Letters*, 28(4):369–371, 1992.

[9] L. Dong, J. L. Archambault, L. Reekie, P. St. J. Russell, and D. N. Payne. Photoinduced absorption change in germanosilicate preforms: evidence for the color-center model of photosensitivity. *Applied Optics*, 34(18):3436–3440, 1995.

[10] M. J. F. Digonnet. A Kramers-Kronig analysis of the absorption change in fiber gratings. *Proceedings of the SPIE*, 2841:109–120, 1996.

[11] P. Russell, D. P. Hand, Y. T. Chow, and L. J. Poyntz Wright. Optically-induced creation, transformation and organisation of defects and colour-centres in optical fibres. *Proceedings of the SPIE*, 1516:47–54, 1991.

[12] M. G. Sceats and P. A. Krug. Photoviscous annealing-dynamics and stability of photorefractivity in optical fibres. *Proceedings of the SPIE*, 2044:113–120, 1993.

[13] J. P. Bernardin and N. M. Lawandy. Dynamics of the formation of Bragg gratings in germanosilicate optical fibers. *Optics Communications*, 79(3–4):194–199, 1990.

[14] B. Poumellec, P. Guenot, I. Riant, P. Sansonetti, P. Niay, P. Bernage, and J. F. Bayon. UV induced densification during Bragg grating inscription in Ge:SiO_2 preforms. *Optical Materials*, 4(4):441–449, 1995.

[15] B. Poumellec, P. Niay, M. Douay, and J. F. Bayon. The UV-induced refractive index grating in Ge:SiO_2 preforms: Additional CW experiments and the macroscopic origin of the change in index. *Journal of Physics D (Applied Physics)*, 29(7):1842–1856, 1996.

[16] D. Wong, S. B. Poole, and M. G. Sceats. Stress relief: proof of the mechanism of photo-induced index change. *Integrated Photonics Research*, pd16:408–412, 1992.

[17] P. Y. Fonjallaz, H. G. Limberger, R. P. Salathe, F. Cochet, and B. Leuenberger. Tension increase correlated to refractive-index change in fibers containing UV-written Bragg gratings. *Optics Letters*, 20(11):1346–1348, 1995.

[18] H. G. Limberger, P. Y. Fonjallaz, R. P. Salathe, and F. Cochet. Compaction- and photoelastic-induced index changes in fiber Bragg gratings. *Applied Physics Letters*, 68(22):3069–3071, 1996.

[19] S. E. A. Bayoumi and E. K. Frankl. Fundamental relations in photo-plasticity. *British Journal of Applied Physics*, 4(10):306–310, 1953.

[20] A. D. Yablon, M. F. Yan, P. Wisk, F. V. DiMarcello, J. W. Fleming, W. A. Reed, E. M. Monberg, D. J. DiGiovanni, J. Jasapara, and M. E. Lines. Refractive index perturbations in optical fibers resulting from frozen-in viscoelasticity. *Applied Physics Letters*, 84(1):19–21, 2004.

[21] P. M. Sutton. Stress measurements in circular cylinders. *Journal of the American Ceramic Society*, 41(3):103–109, 1958.

[22] M. J. Saunders. Determination of the stress in optical fibers by means of a polariscope. *Review of Scientific Instruments*, 47(4):496–500, 1976.

[23] P. L. Chu and T. Whitbread. Measurement of stresses in optical fibre or preform. *Electronics Letters*, 18(1):28–29, 1982.

[24] P. S. Theocaris and E. E. Gdoutos. *Matrix Theory of Photoelasticity*. Springer Series in Optical Sciences, 1979.

[25] T. Abe, Y. Mitsunaga, and H. Koga. Novel measurement method for axial residual stress in optical fibre. *Electronics Letters*, 21(1):4–5, 1985.

[26] A. C. Kak and M. Slaney. *Principles of Computerized Tomographic Imaging*. Society of Industrial and Applied Mathematics SIAM, 2001.

[27] S. Inoué and C. J. Koester. Optimum half-shade angle in polarizing instruments. *Journal of the Optical Society of America*, 49(6):556–559, 1959.

[28] K. W. Raine. A microscope for measuring axial stress profiles in optical fibres with high spatial resolution and low noise. In *OFMC '97. 4th Optical Fibre Measurement Conference*, pages 269–272. Optical Society of America, 1997.

[29] K. W. Raine and A. J. Parker. Measurement of two dimensional axial stress profiles of optical fibres with high spatial resolution. In *Symposium on Optical Fiber Measurements 1998 (NIST SP-930)*, pages 57–60, 1998.

[30] Y. Park, T. J. Ahn, Y. H. Kim, W. T. Han, U. C. Paek, and D. Y. Kim. Measurement method for profiling the residual stress and the strain-optic coefficient of an optical fiber. *Applied Optics*, 41(1):21–26, 2002.

[31] Y. Park, U. C. Paek, and D. Y. Kim. Complete determination of the stress tensor of a polarization-maintaining fiber by photoelastic tomography. *Optics Letters*, 27(14):1217–1219, 2002.

Chapter 2

Residual stresses and strains in optical fibers

Strain is generally defined as the ratio of geometrical change in length to the original length along the coordinates of a body [1, 2]. In the elastic region, a repulsing force, the stress, always accompanies strains. Both stresses and strains basically relax when the body is cut into small slices. In contrast, inelastic strains only appear in the plastic range of deformation. They manifest themselves as an underlying anisotropy of the body, comparable to the anisotropy of crystals. They are not accompanied by stresses and still remain when the body is cut into small slices. Both inelastic and elastic strains are frozen into the fiber during its fabrication process [3]. Elastic strain is introduced into the fiber thermally during the preform fabrication process, but also mechanically during the fiber drawing process. It can be modified when the core glass of the fiber changes its volume, i.e. when it densifies or dilates. The exact measurement of fiber stress changes thus allows to draw conclusions about changes in the density of the core, which are observed for example in UV-irradiated fibers [4]. During fiber drawing, the silica glass is cooled down very rapidly to room temperature under a constant external force. In this viscoelastic regime, both elastic and inelastic strains are added to the fiber.

In the first section of this chapter, we provide a plane strain analysis for stresses introduced into the fiber due to a difference in volume expansion between core and cladding. This analysis has been used traditionally in the literature to describe stresses caused by a radial gradient in thermal expansion coefficient, and is also applied here for the case of radiation-induced compaction of the core glass. We thus obtain a direct linear relationship between densification and core stress increase. The analysis also gives the connection between the different components of the stress tensor for both radial and non-radial symmetry. In both cases, the full stress tensor can basically be determined only from the axial stress profile.

In the second section, elastic and inelastic strains are reviewed that build up when the fiber is in a viscoelastic regime during the drawing process. The drawing-induced elastic strains, i.e. stresses, are superposed to the thermal stresses that develop already during preform fabrication. The total stress is thus the sum of two different stress components. In parallel, inelastic strain is frozen into the fiber, resulting in both, a dilatation and a strain anisotropy.

2.1 Elastic stresses in optical fibers

In the general case of three-dimensional problems of elasticity, six stress components, six strain components as well as three displacement components have to be ascertained [1, 2]. Therefore, the components must satisfy 15 governing equations throughout the body: three equations of equilibrium, six stress-strain relations, and six strain-displacement relations. In addition, the boundary conditions of the respective problem need to be considered. However, for a large quantity of elastic problems, simplifying assumptions can be made that reduce the complexity of the solution considerably. A simplifying assumption is for example the independence of stress or strain from one of the coordinates, in the following taken as z. The three-dimensional problem can thus be reduced to a two-dimensional one [1, 2]. Two classes of these simplified two-dimensional problems exist: plane stress and plane strain. In the case of plane stress, the axial stress σ_{zz} is assumed to vanish, whereas in plane strain problems, the axial strain ϵ_{zz} is assumed to be constant. Plane stress problems apply to bodies where the axial extension is small as compared to the transverse dimensions, i.e. for thin disks. In contrast, plane strain problems are used to solve problems where the axial extension is large as compared to the transverse dimensions, i.e. for long, prismatic bodies. To obtain information about elastic properties in optical fiber, a plane strain problem must thus be solved.

2.1.1 Plane strain analysis

To account for the geometry of the fiber, cylindrical coordinates are used in the following. The fiber is basically made of silica, which is assumed to be an isotropic and linear elastic material. We introduce a free strain ϵ_f in our analysis, that accounts for volume changes within the fiber due to gradients in thermal expansion coefficient or irradiation with laser light. The free strain serves as a source for both resulting stresses and strains. The reduced stress-strain relation, or Hooke's law, then reads [1, 2]

$$\epsilon_{rr} = \frac{1-\nu^2}{E}\left[\sigma_{rr} - \frac{\nu}{1-\nu}\sigma_{\theta\theta}\right] + (1+\nu)\epsilon_f = \frac{1+\nu}{E}(\sigma_{rr} - \sigma_{zz}), \qquad (2.1)$$

$$\epsilon_{\theta\theta} = \frac{1-\nu^2}{E}\left[\sigma_{\theta\theta} - \frac{\nu}{1-\nu}\sigma_{rr}\right] + (1+\nu)\epsilon_f = \frac{1+\nu}{E}(\sigma_{\theta\theta} - \sigma_{zz}), \qquad (2.2)$$

$$\epsilon_{r\theta} = \frac{1+\nu}{E}\sigma_{r\theta}, \qquad (2.3)$$

where E is Young's modulus and ν Poisson's ratio. The axial stress σ_{zz} is determined by

$$\sigma_{zz} = \nu(\sigma_{rr} + \sigma_{\theta\theta}) - E\epsilon_f. \qquad (2.4)$$

The geometry of the deformation and the distribution of strain must be consistent with the preservation of body continuity, i.e. the conditions of compabtibility have to be fullfilled. Therefore, the strains must comply with the compatibility equation [1]

$$\frac{\partial^2 \epsilon_{\theta\theta}}{\partial r^2} + \frac{1}{r^2}\frac{\partial^2 \epsilon_{\theta\theta}}{\partial \theta^2} + \frac{2}{r}\frac{\partial \epsilon_{\theta\theta}}{\partial \theta} - \frac{1}{r}\frac{\partial \epsilon_{rr}}{\partial r} = \frac{1}{r}\frac{\partial^2 \epsilon_{r\theta}}{\partial r \partial \theta} + \frac{1}{r^2}\frac{\partial \epsilon_{r\theta}}{\partial \theta}, \qquad (2.5)$$

where body forces have been neglected. Alternatively, the equation of compatibility can be expressed in terms of stress [1]

$$\Delta\left(\sigma_{rr} + \sigma_{\theta\theta} + \frac{E}{1-\nu}\epsilon_f\right) = 0, \qquad (2.6)$$

where Δ is the Laplace-operator in polar coordinates and is given by

$$\Delta = \frac{\partial^2}{\partial r^2} + \frac{1}{r}\frac{\partial}{\partial r} + \frac{1}{r^2}\frac{\partial^2}{\partial \theta^2}. \qquad (2.7)$$

For the further analysis, we introduced the Airy stress function $\Phi(r,\theta)$ that is related to the three stress components through

$$\sigma_{rr} = \frac{1}{r}\frac{\partial \Phi}{\partial r} + \frac{1}{r^2}\frac{\partial^2 \Phi}{\partial \theta^2} \qquad (2.8)$$

$$\sigma_{\theta\theta} = \frac{\partial^2 \Phi}{\partial r^2} \qquad (2.9)$$

$$\sigma_{r\theta} = \frac{1}{r^2}\frac{\partial \Phi}{\partial \theta} - \frac{1}{r}\frac{\partial^2 \Phi}{\partial r \partial \theta}. \qquad (2.10)$$

The definition of stresses as derivatives of the Airy stress function $\Phi(r,\theta)$ ensures that the equations of equilibrium, satisfying the static equilibrium of forces, are equally fulfilled. In the following, we assume that neither Young's modulus nor Poisson's ratio are functions of position. Then, by inserting the stress-strain relationships into the compatibility equation, and by further substituting the stresses by the corresponding expressions which contain the Airy stress function, one obtains the governing differential equation for plane strain [1, 2]

$$\Delta^2 \Phi = \frac{E}{1-\nu} \Delta \epsilon_f = \Delta \sigma_{zz}. \qquad (2.11)$$

In 1941, Gatewood demonstrated that analytical solutions for the Airy stress function can be obtained from equation (2.11) and presented solutions that depend only on radius as well as on both radius and azimuth angle [5] . The corresponding stresses and strains can then be deduced using equations (2.8) to (2.10). However, for radial symmetric cylinders, the complex calculation of the Airy stress function can be skipped, as will be demonstrated in the following section.

2.1.2 Stresses in cylinders with radial symmetry

The first analytical solutions for thermal stresses in a radial symmetric cylinder have been introduced by Poritsky in 1934 [6]. However, his approach was purely heuristic and thus no derivation from the theory of plane strain presented in the previous section was provided. The basic idea allowing the determination of stresses without the necessity of solving equation (2.11) for the Airy stress function was suggested by O'Rourke in 1941 [7] and fully integrated in the stress analysis of circular cylinders by Sutton in 1958 [8]. The Laplacian is applied to equation (2.4) and the resulting equation is inserted into equation (2.6). As a result, one gets

$$\Delta \sigma_{zz} = \Delta(\sigma_{rr} + \sigma_{\theta\theta}), \qquad (2.12)$$

which can alternatively be written as

$$\sigma_{zz} = \sigma_{rr} + \sigma_{\theta\theta} + \chi, \qquad (2.13)$$

where the harmonic function χ is defined by $\Delta \chi = 0$. For circular symmetry, $\chi = 0$, and equation (2.13) turns out to be the so called "sum rule" (see, e.g. [9]), which connects the three stress components only for the case of circular symmetry. As the "sum rule" is derived from equation (2.11), it applies only to cylinders where Young's modulus and Poisson's ratio do not depend on position [10]. To proceed further, we introduce the equation of equilibrium

$$\sigma_{\theta\theta} = \sigma_{rr} + r \frac{\partial \sigma_{rr}}{\partial r}, \qquad (2.14)$$

which can be obtained from the definition of the stress function (2.8) to (2.10) when circular symmetry is assumed. Inserting the "sum rule" into equation (2.14) yields a differential equation which can be solved for the radial stress [9]

$$\sigma_{rr}(r) = \frac{1}{r^2} \int_0^r \sigma_{zz}(\rho)\rho d\rho. \tag{2.15}$$

Clearly, σ_{rr} must vanish outside the fiber, which is only guaranteed when the integral of the axial stress over the fibers area vanishes, i.e.

$$\int_A \sigma_{zz} dA = 0. \tag{2.16}$$

Equation (2.16) reflects the fact that according to classical mechanics, the resultant force on a free static surface must always vanish. Equation (2.16) is sometimes referred to as "St. Vernant's principle".

Inserting equations (2.4) in equation (2.14) and applying again the "sum rule" as well as the boundary condition $\sigma_{rr}(R) = 0$, where R is the outer radius of the cylinder, one gets the following expressions for axial, radial and hoop stress inside the cylinder [8]:

$$\sigma_{zz}(r) = \frac{E}{1-\nu}\left[\frac{2}{R^2}\int_0^R \epsilon_f(r) r dr - \epsilon_f(r)\right] \tag{2.17}$$

$$\sigma_{rr}(r) = \frac{E}{1-\nu}\left[\frac{1}{R^2}\int_0^R \epsilon_f(r) r dr - \int_0^r \epsilon_f(r) r dr\right] \tag{2.18}$$

$$\sigma_{\theta\theta}(r) = \frac{E}{1-\nu}\left[\frac{1}{R^2}\int_0^R \epsilon_f(r) r dr - \int_0^r \epsilon_f(r) r dr - \epsilon_f(r)\right] \tag{2.19}$$

More sophisticated formulas, that also account for the radial dependence of elastic and thermal properties, have been presented by Scherer in 1979 [9]. If we now assume a simple step-profile of free strain

$$\epsilon_f = \begin{cases} \epsilon_{f,1}, & 0 \leq r < \varrho, \\ \epsilon_{f,2}, & \varrho \leq r < R, \end{cases} \tag{2.20}$$

where ϱ is the radius of the fiber core, the integrals in equations (2.17) to

(2.19) can readily be solved for the core to get

$$\sigma_{zz}^{(1)} = \frac{E}{1-\nu}(\epsilon_{f,1} - \epsilon_{f,2})\left(\frac{\varrho^2}{R^2} - 1\right) \tag{2.21}$$

$$\sigma_{rr}^{(1)} = \frac{1}{2}\sigma_{zz}^{(1)} \tag{2.22}$$

$$\sigma_{\theta\theta}^{(1)} = \frac{1}{2}\sigma_{zz}^{(1)} \tag{2.23}$$

and the corresponding solutions for the cladding

$$\sigma_{zz}^{(2)} = \frac{E}{1-\nu}(\epsilon_{f,1} - \epsilon_{f,2})\frac{\varrho^2}{R^2} \tag{2.24}$$

$$\sigma_{rr}^{(2)} = \frac{E}{2(1-\nu)}(\epsilon_{f,1} - \epsilon_{f,2})\frac{\varrho^2}{R^2}\left(1 - \frac{R^2}{r^2}\right) \tag{2.25}$$

$$\sigma_{\theta\theta}^{(2)} = \frac{E}{2(1-\nu)}(\epsilon_{f,1} - \epsilon_{f,2})\frac{\varrho^2}{R^2}\left(1 + \frac{R^2}{r^2}\right). \tag{2.26}$$

The solutions agree with the more complex solutions presented by Poritsky [6] and Brugger [11], who also account for the difference in elastic properties between core and cladding. Solutions of equations (2.17) to (2.19) for parabolic free strain profiles have been presented by Shibata [12] and Scherer [13].

Krohn mentioned for the first time in 1969 that additional stresses might be introduced in the cladding of a step-profile fiber, when the cladding solidifies first upon cooling and is thus subjected to hydrostatic pressure from the core [14]. Corrected formulas for this effect were presented by Scherer and Krohn in 1980 [15] and extended to parabolic strain profiles by Scherer in 1982 [16]. However, we neglect this effect in the following and use equations (2.21) to (2.26) to describe stresses in circular symmetric cylinders.

2.1.3 Stresses in cylinders of arbitrary cross section

In the general case of cylinders with arbitrary free-strain profiles, no analytical solutions for the four stress components can be found. Solutions presented by Gatewood [5] only apply for a limited number of non radial-symmetric free-strain profiles. However, it has been demonstrated by Puro et Kell [17] that, if the axial stress profile inside a cylinder is known, the Airy stress function and thus all other components of the stress tensor can be determined. This result is of particular interest, because most experimental setups allow the acquisition of axial stress profiles, whereas no information about the other stress components is obtained (cf. Chapter 4).

The starting point of the method is the differential equation (2.11). The equation can be integrated to get

$$\Delta \Phi = \sigma_{zz} - \chi, \qquad (2.27)$$

where χ is an arbitrary harmonic function, i. e. $\Delta \chi = 0$. χ has to be chosen according to the boundary conditions of the problem. In a next step, we expand the Airy stress-function as well as the axial stress profile into a Fourier series

$$\Phi(r,\theta) = \sum_n f_n(r) \exp(j n\theta) \qquad (2.28)$$

$$\sigma_{zz}(r,\theta) = \sum_n s_n(r) \exp(j n\theta) \qquad (2.29)$$

and choose the ansatz

$$\chi(r,\theta) = \sum_n k_n r^n \exp(j n\theta) \qquad (2.30)$$

for the harmonic function χ [18]. Here, we use the complex notation with $j = \sqrt{-1}$ to express the Fourier series instead of the real expressions used in [17] and [19], as it allows a more elegant description of the solution. The Fourier-coefficients k_n can be determined from Green's theorem [18] and the boundary conditions [19] to be

$$k_n = 2(n+1) \int_0^1 s_n \, \rho^{n+1} \, d\rho. \qquad (2.31)$$

Equation (2.27) is rewritten in terms of the Fourier coefficients and yields

$$\left(\frac{\partial^2}{\partial r^2} + \frac{1}{r} \frac{\partial}{\partial r} - \frac{n^2}{r^2} \right) f_n = s_n - k_n r^n = g_n, \qquad (2.32)$$

where g_n has been introduced for simplification. The boundary conditions for a cylinder of infinite length are given by

$$F \, |_{r=1} = 0; \qquad \frac{\partial F}{\partial r} \, |_{r=1} = 0, \qquad (2.33)$$

where the radius of the cylinder has been normalized to unity. The corresponding boundary conditions in the Fourier domain are

$$f_n \, |_{r=1} = 0; \qquad \frac{\partial f_n}{\partial r} \, |_{r=1} = 0. \qquad (2.34)$$

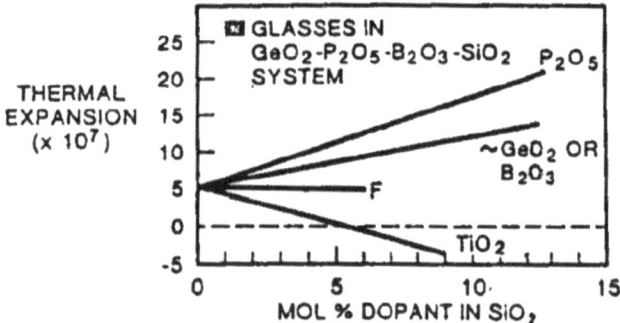

Figure 2.1: *Dependence of the thermal expansion coefficient on dopant concentration for the most commonly used core dopants ([20]).*

The solution of the boundary-value problem given by equations (2.32) and (2.34) is rather complex and has been derived in detail in the appendix of [19]. The solution is

$$f_n = -\frac{1}{2n}\left[r^n \int_r^1 g_n \rho^{-n+1} d\rho + r^{-n} \int_0^r g_n \rho^{n+1} d\rho\right]. \qquad (2.35)$$

When the axial stress profile of a fiber has been determined experimentally, than it's Fourier coefficients s_n are calculated according to equation (2.29) using a fast Fourier transform algorithm. These coefficients are the starting point to obtain the coefficients k_n with equation (2.31) and further g_n with equation (2.32). Subsequently, the Fourier coefficients of the Airy stress function are determined with equation (2.35). To get the complete stress tensor, the Airy stress-function given by equation (2.28) with Fourier coefficients according to (2.35) has to be inserted into equations (2.8) to (2.10).

2.1.4 Thermal stresses

Thermal stresses are introduced into the fiber when cladding and core solidify, i.e. when the temperature in the fiber is smaller than the softening temperature T_S of the glass. The origin of the thermal stress is a free strain that results from a mismatch in thermal expansion coefficients between different regions of the fiber, most commonly between the core and the cladding. The free strain due to thermal mismatch is given by

$$\epsilon_{f,i} = \int_{T_{S,i}}^{T_R} \alpha_i(T) dT \qquad (2.36)$$

where α_i is the thermal expansion coefficient of the respective region i, T_R is room temperature, and $T_{S,i}$ is the virtual temperature of the region i where stress starts to develop upon cooling to room temperature [21]. In general, the virtual temperature is assumed to be identical to the glass transition temperature, which is defined as the temperature where the viscosity of the glass equals 10^{13} poise [22]. The glass transition temperature is different for optical preforms and fibers, as the latter are cooled much faster [23]. For preforms, values of about 1100 - 1200 °C are assumed, whereas temperatures up to 2000 °C might be reached for optical fibers [24]. The transition temperature might differ considerably between the deposition tube and the deposited layers for fiber preforms fabricated with the Modified Chemical Vapor Deposition (MCVD) process [25]. In contrast, a distinction between different regions can be omitted for preforms fabricated using Outside Vapor Deposition (OVD) or Vapor-axial deposition (VAD) technologies.

In general, the thermal expansion coefficient is a function of temperature [26]. For germanium-doped fibers, however, satisfactory agreement between theory and experiment can be found if the thermal expansion coefficient is assumed to be constant [26], so that equation (2.36) is simplified to

$$\epsilon_{f,i} = -\alpha_i(T_{S,i} - T_R). \tag{2.37}$$

In Fig. 2.1, the thermal expansion coefficient at room temperature is illustrated for the most important fiber dopants as a function of dopant concentration [20]. The addition of germanium, phosphorus and boron increases the expansion coefficient with respect to pure silica, whereas fluor and titanium decrease it. For the most commonly used fiber dopant, germanium, the difference in free strain between core and cladding, $\epsilon_{f,1} - \epsilon_{f,2}$ is of negative sign. According to equation (2.21), the corresponding axial core stress is then positive, and increases with decreasing core diameter. The axial cladding stress is slightly negative and approaches zero for decreasing core diameters.

2.1.5 UV-induced stresses

When the core of a germanium-doped optical fiber is irradiated with UV-laser light, its initial stress value has been found to increase [27]. The increase in core stress has been attributed to a densification of the doped core glass [4], which introduces a free strain and thus reweighs the fibers stress profile. The free strain thus accounts for an unconstrained densification or volume change, which provokes an elastic response, as the core is surrounded by the silica cladding. The measurement of core stress changes in an optical fiber thus offers the possibility to gather information about unconstrained density or volume changes introduced by irradiation. Radiation-induced core stress

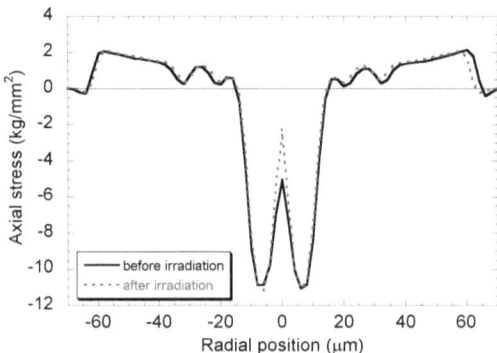

Figure 2.2: *Core stress changes in a phosphorus-doped fiber drawn with 225 g due to irradiation with an ArF excimer laser. The core stress increases due to a compaction of the core glass.*

changes are illustrated for a hydrogen-loaded phosphorus-doped fiber irradiated with an ArF excimer-laser in Fig. 2.2. The fiber has been manufactured at FORC[1] and has a fluorine and phosphorus-codoped inner cladding.

Gusarov et al. [28] and Kherbouche et al. [10] both presented solutions for the dependence of axial stress changes on a periodic core densification as expected to occur in fiber Bragg gratings. However, the results of [10] are derived only numerically using a finite element method, and the analytical method given by [28] is rather complex. Here, we interpret the densification as a free strain and exploit formula (2.21) to derive a simple relationship between axial stress change and unconstrained core densification. The obtained formula does not apply to periodic core densification as the solutions presented in [28] and [10]; however, as the resolution of common stress measurement setups in general only allow to gather information about mean stress changes, the approximation is completely sufficient for the purpose of this work.

The axial core stress difference with free strain is thus found using equation (2.21) to be

$$\Delta\sigma_{zz} = \frac{E}{1-\nu}\epsilon_{f,UV}\left(\frac{\varrho^2}{R^2} - 1\right), \qquad (2.38)$$

where $\epsilon_{f,UV}$ is the UV-induced free strain in the core. For small elongations,

---

[1]Fiber Optics Research Center at the General Physics Institute of the Russian Academy of Science, Moscow, Russia.

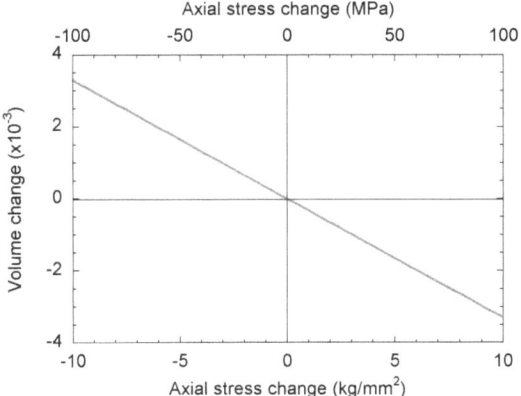

Figure 2.3: *Axial core stress changes as a function of core volume changes. The axial core stress increases with decreasing volume, i.e. for compaction.*

the strain is related to volume changes by the approximation

$$\epsilon_f = \frac{\Delta L}{L} \approx \frac{1}{3}\frac{\Delta V}{V_0}, \qquad (2.39)$$

which can be included in equation (2.38) to get the dependence of volume change on axial stress change

$$\frac{\Delta V}{V_0} = 3\,\frac{(1-\nu)}{E}\left(\frac{\varrho^2}{R^2}-1\right)^{-1}\Delta\sigma_{zz} \approx -3\,\frac{(1-\nu)}{E}\Delta\sigma_{zz}. \qquad (2.40)$$

The approximation in equation (2.40) holds for optical fibers, where the squared ratio between core and cladding radius is assumed to vanish, as the core diameter is in general more than one order of magnitude smaller than the cladding diameter. Equation (2.40) is essentially similar to the stress-densification dependency given by Schenker *et al.* in [29] without providing any derivation. In Fig. 2.3, equation (2.40) is illustrated graphically. An increase in axial core stress is accompanied by a negative change in core volume, i.e. by a compaction of the core glass. In contrast, negative stress changes indicate a dilatation of the core. The axial stress in Fig. 2.3 is given in Pascals as well as in kg/mm². Both units can be found in the literature and are linked by the acceleration of gravity, so that 10 MPa correspond to about 1 kg/mm². In the following, the stress in the calculations are assumed to be given in Pascals, whereas most figures illustrate stress in kg/mm², which is consistent with the notation used in our earlier work.

2.2 Drawing-induced stresses and strains: Viscoelasticity

In the preceding section, general expressions for the calculation of stresses in fully elastic cylindrical bodies have been presented. The generation of stress is triggered by the presence of a free strain field $\epsilon_f(r,\theta)$, which can be introduced thermally as well as by irradiation. In this section, we present additional stresses and strains that are introduced in the fiber during its drawing process, when the fiber is still in the viscoelastic regime. In the first subsection, we present drawing-induced stresses that build up in the fiber due to a mismatch in viscosity between the core and the cladding material. The different viscosities result in different amounts of elastic strain in the core and cladding during fiber drawing, that convert to a different amount of stress when the drawing force is released. In the second subsection, we illustrate the generation of frozen-in strains in optical fibers. These strains are inelastic, i.e. they are not accompanied by the generation of stress. They are build up due to a quenching of the relaxational part of the fibers elastic properties.

2.2.1 Drawing-induced stresses

As the axial components of stress and strain are large in comparison to the transversal components during the fiber drawing process, the drawing-induced stress generation can be considered as a one-dimensional viscoelastic problem [23]. Drawing-induced stress can be calculated for fibers with an arbitrary number of core and cladding regions [24, 26]. However, in this section, we restrict our analysis to a fiber consisting of only one core and one cladding region.

During fiber drawing, different amounts of axial stress are built up in the core and in the cladding, respectively, due to a mismatch in the viscoelastic properties. The dependence of axial stress on drawing tension in the initial, viscoelastic part of the drawing process is given by [24]

$$\sigma_{zz}^{(i),\text{init}} = \frac{\eta_i}{\sum_i \eta_i A_i} \tau, \qquad i = 1, 2, \tag{2.41}$$

where τ is the fiber drawing tension in Pa, η_i is the viscosity of the region of area A_i, $i = 1$ refers to the core and $i = 2$ to the cladding. The drawing tension is given by the drawing force divided by the fiber's cross sectional area, $\tau = F/A$ with $A = A_1 + A_2$. During fiber drawing, the mechanical equilibrium requires

$$A_1 \sigma_{zz}^{(1),\text{init}} + A_2 \sigma_{zz}^{(2),\text{init}} = \tau, \tag{2.42}$$

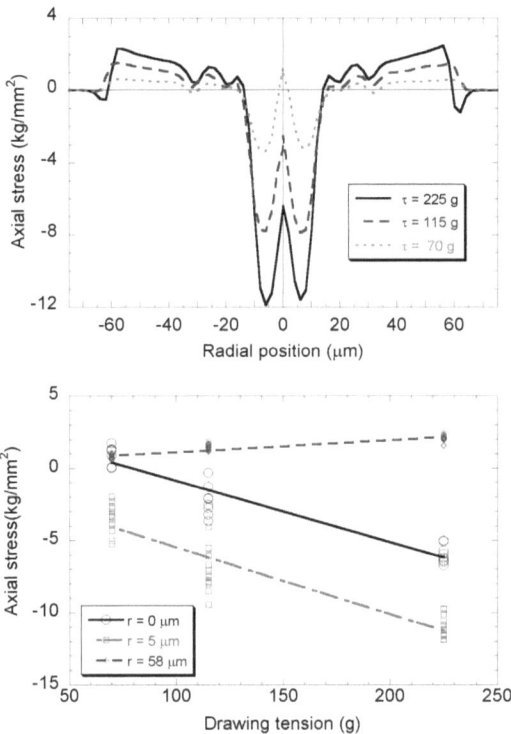

Figure 2.4: (a) Stress profiles of optical fibers drawn from the same phosphorus-doped preform at different drawing tensions. (b) Linear dependence of stress on drawing tension at three different positions in the phosphorus-doped fiber.

in accordance with (2.41); after tension release, equation (2.42) turns into

$$A_1 \sigma_{zz}^{(1),\text{final}} + A_2 \sigma_{zz}^{(2),\text{final}} = 0. \tag{2.43}$$

It is assumed that the elastic part of the total deformation is the same for all regions of the fiber, i.e. the difference in elastic strain before and after tension release is constant in both the core and the cladding, i.e.

$$\epsilon_{zz}^{(1),\text{final}} - \epsilon_{zz}^{(1),\text{init}} = \epsilon_{zz}^{(2),\text{final}} - \epsilon_{zz}^{(2),\text{init}}. \tag{2.44}$$

Equations (2.41) to (2.44) can readily be solved to get the final drawing induced axial stresses in the core and in the cladding, respectively,

$$\sigma_{zz}^{(1),\text{final}} = \left(\frac{\eta_1}{A_1\eta_1 + A_2\eta_2} - \frac{E_1}{A_1E_1 + A_2E_2}\right)\tau = \tilde{m}_{\text{Co}}\,\tau, \qquad (2.45)$$

$$\sigma_{zz}^{(2),\text{final}} = \left(\frac{\eta_2}{A_1\eta_1 + A_2\eta_2} - \frac{E_2}{A_1E_1 + A_2E_2}\right)\tau = \tilde{m}_{\text{Cl}}\,\tau, \qquad (2.46)$$

where $\tilde{m}_{\text{Co}}$ and $\tilde{m}_{\text{Cl}}$ have been introduced for simplification. According to equations (2.45) to (2.46), the drawing-induced stresses depend linearly on the drawing tension, as shown for the first time by Bachmann et al. in [26]. The higher the drawing tension, the higher is the compressive core stress and the tensile cladding stress introduced during fiber drawing. In Fig. 2.4 (a), stress changes as a function of drawing tension are illustrated for a phosphorus-doped fiber manufactured at FORC. As expected, both core and inner cladding stresses decrease with drawing tension, whereas the stress in the undoped cladding tube increases. The linear dependence of stress on drawing tension is depicted for this fiber in Fig. 2.4 (b) in the core, the inner cladding and the cladding tube. Extrapolation of the stress values to zero drawing tension yields the stress level in the preform, i.e. the thermally induced stresses. The thermal core stresses are found to be positive, which is in agreement with equations (2.21) and (2.37). As the fiber presented in Fig. 2.4 consists of three regions with different viscoelastic properties, the simple model presented above must in general be extended to apply to the problem. However, we restrict our analysis here to a simple fiber consisting of two regions, as we are more interested to show the principal qualitative behavior.

2.2.2 Inelastic strains in optical fibers

In addition to the stresses introduced during fiber-drawing, also inelastic strains, i.e. glass deformations that are not accompanied by stresses, are generated in the fiber. This phenomenon has only been discoveerd recently by Yablon et al. [30]. The inelastic strains are caused by the large temperature changes during fiber-drawing, that freeze-in the relaxational contribution of the elastic properties, i.e. the compressibility and the shear compliance. The inelastic strains mainly occur in the fiber region of highest viscosity, which in general is the fiber cladding. For the modeling of the effect, however, we assume a homogeneous glass rod and uniaxial stress. The uniaxial stress τ is constant over the fiber radius and equals the drawing tension. The drawing force is sometimes given as a weight in grams and can be converted into a force by multiplication with the acceleration of gravity.

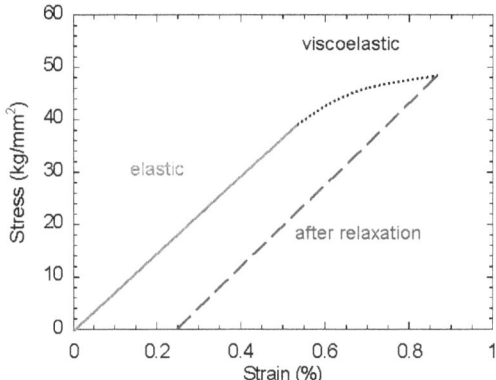

Figure 2.5: During fiber draw, a constant tension is applied to the fiber, that results in an elastic strain (red line) and an inelastic strain or creep (black curve). The inelastic strain is frozen into the fiber to its value at the fictive temperature. When the drawing tension is released (blue line), the inelastic strain remains permanently in the fiber.

Linear elasticity relates the drawing tension to the axial and tangential strain components by

$$\epsilon_{zz} = \frac{1}{E}\,\tau, \tag{2.47}$$

$$\epsilon_\perp = -\frac{\nu}{E}\,\tau. \tag{2.48}$$

Equations (2.47) and (2.48) can be expressed as an isotropic dilatation and an anisotropic component by

$$\theta = \epsilon_{zz} + 2\,\epsilon_\perp = K(t)\,\frac{\tau}{3}, \tag{2.49}$$

$$\Delta\epsilon = \epsilon_{zz} - \epsilon_\perp = J(t)\,\frac{\tau}{2}, \tag{2.50}$$

where the compressibility $K = 3(1 - 2\nu)/E$ and the shear compliance $J = 2(1+\nu)/E$ have been introduced. Both compressibility and shear compliance are a function of time t in the viscoelastic regime. There, both parameters consist of an instantaneous elastic part and a relaxational viscoelastic part, which relaxes with a characteristic retardation time. This behavior is illustrated qualitatively in Fig. 2.5. During fiber draw, a constant stress is applied to the fiber, which results in an instantaneous elastic strain (solid line) and a relaxational time-dependent creep, i.e. inelastic strain (dotted line). When the fiber is quenched during the drawing-process, the temperature rapidly decreases as a function of time [23]. As a result, the inelastic parts of compressibility and shear compliance cannot respond in the available time and are "frozen" into the glass. It was reported [30] that these inelastic contributions are frozen to their respective values at the glass fictive temperature T_{fict}. The fictive temperature of a glass is defined as the temperature at which the supercooled glass forming liquid would find itself in equilibrium [31]. Equations (2.49) and (2.50) turn into

$$\theta(T_{\text{room}}) = [K_{\text{rel}}(T_{\text{fict}}) + K_{\text{el}}(T_{\text{room}})] \frac{\tau}{3}, \qquad (2.51)$$

$$\Delta\epsilon(T_{\text{room}}) = [J_{\text{rel}}(T_{\text{fict}}) + J_{\text{el}}(T_{\text{room}})] \frac{\tau}{2}, \qquad (2.52)$$

at room temperature T_{room}. When the drawing tension is released, the elastic contributions in equations (2.51) and (2.52) vanish and both dilatation and strain anisotropy only depend on the frozen-in relaxational part of the viscoelastic response. This is illustrated by the dashed line in Fig. 2.5. The relaxational part of the compressibility $K_{\text{rel}}(T_{\text{fict}})$ has been determined experimentally as $0.057\,\text{GPa}^{-1}$ in [32] and can directly be inserted into equation (2.51). The determination of the shear compliance $J_{\text{rel}}(T_{\text{fict}})$, however, is less straightforward and can be found in detail in reference [30]. The final expressions for the dilatation and strain anisotropy at room temperature are given by

$$\theta = 0.019 \text{ GPa}^{-1} \times \tau, \qquad (2.53)$$

$$\Delta\epsilon = 0.015 \left[b \frac{\Gamma(2/b)}{\Gamma^2(1/b)} - 1.06 \right] \text{GPa}^{-1} \times \tau, \qquad (2.54)$$

where Γ is the gamma function and b is the exponent of the Kohlrausch-Williams-Watt or stretched exponential function

$$\Psi(t) = \exp(-(t/\theta)^b), \qquad 0 < b < 1, \qquad (2.55)$$

which is commonly used as an empirical representation of glass relaxation functions [33]. For low-hydroxyl silica, reasonable values of b are within the range $0.5 < b < 1$ [34, 35], which allows to estimate the range of strain anisotropy given by equation (2.54) to be $-0.9 \times 10^{-3} < \Delta\epsilon$ GPa$/\tau < 2.9 \times 10^{-2}$. The strain anisotropy changes by more than one order of magnitude for the possible range of b-values. As b is the only free parameter in equation (2.54), the determination of strain anisotropy as a function of fiber drawing tension thus allows to gather information about the relaxational behavior of the fibers material.

2.3 Conclusions

The origins and mathematical descriptions of elastic strain and stress as well as inelastic strain in optical fibers have been presented. Thermal stresses are generated during the preform fabrication process due to a mismatch in thermal expansion coefficient between fiber core and cladding. During the subsequent drawing process, when the fiber is in a viscoelastic state, both elastic strain with responding stress and inelastic strain are generated. The elastic strain originates from a mismatch in viscoelastic properties between core and cladding, whereas the inelastic strain stems from the quenching of the relaxational part of the viscoelastic properties due to the large temperature gradient involved in fiber drawing. The measurement of the strain anisotropy as a function of drawing tension allows to determine the exponent b of the stretched exponential function which describes the relaxational behavior of the fiber glass. Finally, densification of the core glass due to laser-irradiation creates additional stresses, that can be modeled in analogy to thermal stresses.

Bibliography

[1] A. C. Ugural and S. K. Fenster. *Advanced strength and Applied Elasticity*. Elsevier, 1987.

[2] S. P. Timoshenko and J. N. Goodier. *Theory of Elasticity*. McGraw-Hill, 1970.

[3] A. D. Yablon. Optical and mechanical effects of frozen-in stresses and strains in optical fibers. *IEEE Journal of Selected Topics in Quantum Electronics*, 10(2):300–311, 2004.

[4] H. G. Limberger, P. Y. Fonjallaz, R. P. Salathe, and F. Cochet. Compaction- and photoelastic-induced index changes in fiber Bragg gratings. *Applied Physics Letters*, 68(22):3069–3071, 1996.

[5] B. E. Gatewood. Thermal stresses in long cylindrical bodies. *Philosophical Magazine*, 32:282–301, 1941.

[6] H. Poritsky. Analysis of thermal stresses in sealed cylinders and the effect of viscous flow during annealing. *Physics*, 5:406–411, 1934.

[7] R. C. O'Rourke. Three-dimensional photoelasticity. *Journal of Applied Physics*, 22(7):872–878, 1941.

[8] P. M. Sutton. Stress measurements in circular cylinders. *Journal of the American Ceramic Society*, 41(3):103–109, 1958.

[9] G. W. Scherer. Thermal stresses in a cylinder: Application to optical waveguide blanks. *Journal of Non-Crystalline Solids*, 34(2):223–238, 1979.

[10] F. Kherbouche and B. Poumellec. UV-induced stress field during Bragg grating inscription in optical fibres. *Journal of Optics A: Pure and Applied Optics*, 3(6):429–439, 2001.

[11] K. Brugger. Effect of thermal stress on refractive index in clad fibers. *Applied Optics*, 10(2):437–438, 1971.

[12] N. Shibata, K. Jinguji, M. Kawachi, and T. Edahiro. Nondestructive structure measurement of optical-fiber preforms with photoelastic effect. *Japanese Journal of Applied Physics*, 18(7):1267–1273, 1979.

[13] G. W. Scherer. Stress-induced index profile distortion in optical waveguides. *Applied Optics*, 19(12):2000–2006, 1980.

[14] D. A. Krohn and A. R. Cooper. Strengthening of glass fibers. I. Cladding. *Journal of the American Ceramic Society*, 52(12):661–664, 1969.

[15] G. W. Scherer and A. R. Cooper. Thermal stresses in clad-glass fibers. *Journal of the American Ceramic Society*, 63(5–6):346–347, 1980.

[16] G. W. Scherer. Thermal stresses in optical fibers: fluid core assumption. *Journal of Non-Crystalline Solids*, 51(3):323–332, 1982.

[17] A. E. Puro and K. J. Kell. Complete determination of stress in fiber preforms of arbitrary cross section. *Journal of Lightwave Technology*, 10(8):1010–1014, 1992.

[18] I. N. Bronstein and K. A. Semendjajew. *Taschenbuch der Mathematik*. Verlag Harri Deutsch, 1991.

[19] Y. Park. *Stress measurement of an optical fiber*. PhD thesis, Kwangju Institute of Science and Technology, 2002.

[20] S. R. Nagel. Silica-based glass optical fiber properties and fabrication methods. *Proceedings of the SPIE*, 1085:56–76, 1990.

[21] P. K. Bachmann, W. Hermann, H. Wehr, and D. U. Wiechert. Stress in optical waveguides. I. Preforms. *Applied Optics*, 25(7):1093–1098, 1986.

[22] R. H. Doremus. Viscosity of silica. *Journal of Applied Physics*, 92(12):7619–7629, 2002.

[23] U. C. Paek and C. R. Kurkjian. Calculation of cooling rate and induced stresses in drawing of optical fibers. *Journal of the American Ceramic Society*, 58(7–8):330–335, 1975.

[24] Y. Park, K. Oh, U. C. Paek, D. Y. Kim, and C. R. Kurkjian. Residual stresses in a doubly clad fiber with depressed inner cladding (DIC). *Journal of Lightwave Technology*, 17(10):1823–1834, 1999.

[25] F. Cochet. personal communication.

[26] P. K. Bachmann, D. U. Wiechert, and T. P. M. Meeuwsen. Thermal expansion coefficients of doped and undoped silica prepared by means of PCVD. *Journal of Materials Science*, 23(7):2584–2588, 1988.

[27] P. Y. Fonjallaz, H. G. Limberger, R. P. Salathe, F. Cochet, and B. Leuenberger. Tension increase correlated to refractive-index change in fibers containing UV-written Bragg gratings. *Optics Letters*, 20(11):1346–1348, 1995.

[28] A. I. Gusarov, D. B. Doyle, F. Berghmans, and O. Deparis. Analysis of photoinduced stress distribution in fiber Bragg gratings. *Optics Letters*, 24(19):1334–1336, 1999.

[29] R. E. Schenker and W. G. Oldham. Ultraviolet-induced densification in fused silica. *Journal of Applied Physics*, 82(3):1065–1071, 1997.

[30] A. D. Yablon, M. F. Yan, P. Wisk, F. V. DiMarcello, J. W. Fleming, W. A. Reed, E. M. Monberg, D. J. DiGiovanni, J. Jasapara, and M. E. Lines. Refractive index perturbations in optical fibers resulting from frozen-in viscoelasticity. *Applied Physics Letters*, 84(1):19–21, 2004.

[31] A. Agarwal, K. M. Davis, and M. Tomozawa. A simple IR spectroscopic method for determining fictive temperature of silica glasses. *Journal of Non-Crystalline Solids*, 185:191–198, 1995.

[32] D. M. Krol, K. B. Lyons, S. A. Brawer, and C. R. Kurkjian. High-temperature light scattering and the glass transition in vitreous silica. *Physical Review B*, 33(6):4196–4202, 1986.

[33] G. W. Scherer. Theories of relaxation. *Journal of Non-Crystalline Solids*, 123:75–89, 1990.

[34] R. Böhmer, K. L. Ngai, C. A. Angell, and D. J. Plazek. Nonexponential relaxations in strong and fragile glass formers. *Journal of Chemical Physics*, 99(5):4201–4209, 1993.

[35] M. D. Ediger, C. A. Angell, and S. N. Nagel. Supercooled liquids and glasses. *Journal of Physical Chemistry*, 100:13200–13212, 1996.

Chapter 3

Photosensitivity of optical waveguides

Although the phenomenon of photosensitivity in optical fibers has been discovered already more than a quarter of a century ago by Hill and coworkers [1], there is still no complete consistent physical explanation for it at present. However, it has become more and more evident that several different mechanisms contribute to the overall effect. Laser-irradiation is known to alter chemical properties as defect center concentration [2, 3] as well as structural properties as the stress profile [4] of optical fibers. The latter is attributed to a compaction of the fibers core glass [5], whereas the former is a result of electron migration from different defect sites in the glass matrix. Both phenomena contribute to the laser-induced index change.

The different mechanisms contributing to photosensitivity strongly depend on the specific parameters of the fiber as well as on the UV-illumination conditions. Fiber parameters to be mentioned are not only fiber core dopants, but also the atmosphere composition during the preform fabrication process. Fiber drawing conditions define the final amount of internal stresses [6] as well as fiber defect concentration [7]. Hydrogen-loading of the fiber enhances the photosensitivity considerably [8]. Most important parameters of the UV-source with impact on the photosensitive response are intensity (cw-laser), pulse duration and peak power (pulsed laser), total fluence and wavelength [9].

In the first section of this chapter, the origins of photosensitivity are briefly discussed. The color-center model as well as the compaction model are introduced and the contribution of the two models to the absorption spectrum are demonstrated.

As this work basically deals with compaction-induced index changes in optical fiber, we give a more detailed overview about stress and strain induced index changes, i.e. photoelasticity and photoplasticity, in the second section.

In the elastic regime, strain is linearly related to stress and the descriptions of index change in terms of stress or strain are equivalent. In the viscoelastic regime, however, non-elastic strains can appear that are not accompanied by stress but nevertheless contribute to a modification of the refractive index tensor. The description of index changes in terms of strain appears to be valid for both the elastic and the viscoelastic regime and is thus used throughout this work due to its higher versatility.

In the third section, the photoelastic and photoplastic relationships are used to determine the net core index change triggered by the core glass compaction. In the elastic regime, strain is always accompanied by stresses that can be found by applying the laws of elasticity as derived in section 2.1 for the case of optical fibers. The total or net index change thus consists of a compaction-induced and a stress-induced part.

Different experimental methods have been presented to determine the photosensitive refractive index change as a function of irradiation conditions for a given optical fiber. In the fourth section of this chapter, we give a short overview about the different methods and present in more detail two methods used within this work that allow to get information about index changes.

3.1 Origins of photosensitivity

When the refractive index of a dielectric material is known, its absorption spectrum is also defined and vice versa. This linkage between absorption and index is expressed by the Kramers-Kronig relationship

$$n(\lambda') - 1 = \frac{1}{2\pi} P_C \int_0^\infty \frac{\alpha(\lambda)}{1 - (\frac{\lambda}{\lambda'})^2} d\lambda, \tag{3.1}$$

where P_C is the Cauchy principal value of the integral, $\alpha(\lambda)$ the absorption spectrum, and λ' the observed wavelength. When the absorption spectrum is changed by $\Delta\alpha_i(\lambda)$ in i discrete wavelength intervals ranging from λ_i^- to λ_i^+ by $\Delta\alpha_i(\lambda)$, the corresponding refractive index change yields

$$\Delta n(\lambda') = \frac{1}{2\pi} \sum_i \int_{\lambda_i^-}^{\lambda_i^+} \frac{\Delta\alpha_i(\lambda)}{1 - \left(\frac{\lambda}{\lambda'}\right)^2} d\lambda. \tag{3.2}$$

The position of the different absorption changes $\Delta\alpha_i$ can be determined by optical spectroscopy. Correlation of absorption changes with complementary spectroscopic data allows the identification of different contributions to the overall index change.

3.1.1 Color-center model

By comparison with Electron Spin Resonance [10], cathodoluminescence [11], photoluminescence [12] and other spectroscopic-data, a variety of absorption changes in the UV could be assigned to different defects, the so-called color centers, in the glass matrix. A defect is defined as a deviation of the fourfold SiO_4 bonding configuration by the rupture of a bond or by the replacement of an adjacent atom. Defects are introduced into the fiber already during the preform fabrication [13] and further during fiber draw [7]. Irradiation of the fiber leads to a redistribution of the different absorption bands in the UV; some bands decrease in strength, whereas others increase [2, 3, 14]. In many cases, the corresponding refractive index changes found with equation (3.2) were found to agree satisfactorily with the experimentally measured index changes [15–17]. However, other studies report that color-center population changes alone are insufficient to explain the overall radiation-induced index change [14, 18]. It was thus concluded that at least one more effect must contribute to the phenomenon of photosensitivity.

3.1.2 Compaction model

In 1990, Bernardin and Lawandy were the first to suggest a structural alteration of the core glass after UV-irradiation as a second complementary effect to explain photosensitivity [19]. They proposed the collapse of higher-order ring structures to result in a densification of the glass. As shown in more detail in the next section, the Lorentz-Lorenz law predicts an almost linear dependence of refractive index on glass density. Experimental evidence for densification in UV-exposed doped silica was found by scanning the surface of preform samples with atomic force microscopy [20] and interferometric profilometry [21].

As demonstrated in the previous chapter, the densification of the core glass is always accompanied by a modification of the fibers stress profile. A modification of stresses however leads to a change in refractive index due to the photo-elastic effect, as will be shown in more detail in the following section. Wong *et al.* suggested the relief of tensile core stress as main reason for index increase in germanium doped fiber [18, 22]. However, Fonjallaz *et al.* refuted the stress relief model by observing an increase in core stress independent of the initial stress state (tensile or compressive) for germanium doped fibers after UV-illumination [4]. As an increase in core stress corresponds to a decrease in refractive index, it was concluded that the densification and color center induced index change must exceed the photo-elastic one to obtain an overall positive index increase [5]. Elastic and inelastic changes in the glass thus contribute in opposing directions to the absolute value of index change.

Density changes of glass result in a modification of its absorption spectrum [23], and through equation (3.2), to a corresponding change in refractive index. When fused silica densifies, the UV-absorption edge [23, 24] as well as absorption peaks in the infrared [25] shift to longer wavelength. For germanosilicate glass, a significant shift and a broadening of absorption bands related to oxygen deficient centers has been reported in [26].

In this work, however, we exploit the linear relationship between core stress and compaction (equation (2.40)) to get direct information about radiation-induced core volume changes. The fiber cladding forces the irradiated core to give a stress response to the irradiation by constraining it and the stress response can directly be assigned to a volume change.

3.2 Photoelasticity and Photoplasticity

3.2.1 Historical overview

The exact physical principle relating both mean index changes and birefringence to the stress as well as the strain fields in a viscoelastic material remains a matter of discussion almost since the qualitative discovery of photoelasticity by Brewster in 1815 [27]. In 1841, Neumann presented a first qualitative formulation of the relationship between cause and effect in terms of strain [28]. Some twelve years later Maxwell produced a theory in which birefringence was related to the principal stresses [29]. The two theories, however, produced relations of precisely similar form, and they where equally applicable to an isotropic medium strained within the linear elastic regime. Historically, the description of birefringence in terms of stress developed to be the more canonical one, which is also reflected by the general usage of the term "photoelasticity" in the description of the phenomenon.

However, several researchers were aware of the limitation of the theory to perfectly elastic materials and to the linear range of their deformation only. Coker and Filon, who constituted a large contribution to the application of photoelasticity to two-dimensional engineering problems, were the first to suggest for birefringence in the non-linear deformation an equation embracing both strain and stress [30]. Alternatively, the total strain can be subdivided in an elastic contribution and an inelastic part. The elastic strain is accompanied by stresses that can be found by applying the laws of elasticity as derived in section 2.1 for the case of optical fibers. Typical examples for inelastic strains are the core glass compaction of an optical fiber through laser-irradiation as well as drawing-induced inelastic strains as described in section 2.2.2.

In 1938, Mueller developed a theory of photoelasticity essentially based on the strain field in the material. Bayoumi and Frankl presented fundamental relations of photoplasticity essentially based upon stress and inelastic strains [31]. As stress can be converted to elastic strain by applying Hooke's law, an equivalent formulation only in terms of strains can be found. We thus decided to interpret the observed index changes within this work purely in terms of strain. A means to distinguish between elastic and inelastic strains is presented later in section 4.2. An extended overview about the field of photoplasticity is given in [32].

3.2.2 Stress- and strain-induced refractive index changes

For an isotropic material, the relation between strain and the resulting change of refractive index can be expressed in cartesian coordinates by [28, 33]

$$\Delta n_{ii} = n_{ii} - n_0 = \frac{n_0^3}{2}\left[(p_{12} - p_{11})\epsilon_{ii} - p_{12}\theta\right]; \qquad (i = x, y, z) \qquad (3.3)$$

where n_0 is the refractive index of the isotropic material, p_{11} and p_{12} are the Pockels-coefficients, and θ is the mean normal strain (dilatation) given by (2.49). In the elastic regime, the strains in equation (3.3) can be directly related to the corresponding stresses by Hooke's law [34, 35]

$$\epsilon_{ii} = \frac{1}{E}\left[(1+\nu)\sigma_{ii} - \nu\sum_i \sigma_{ii}\right], \qquad (i = x, y, z). \qquad (3.4)$$

Using the stress-strain relationship (3.4), equation (3.3) can be rewritten in terms of stress and reads

$$\Delta n_{ii} = n_{ii} - n_0 = (C_2 - C_1)\sigma_{ii} - C_2\sum_i \sigma_{ii}, \qquad (i = x, y, z), \qquad (3.5)$$

where C_1 and C_2 are the photoelastic constants. As already mentioned above, equation (3.5) only holds in the elastic regime, where elastic strains are always accompanied by stresses, whereas we assume equation (3.3) to be valid for both elastic and inelastic strains [30, 31].

By comparison of equations (3.3) and (3.5), the Pockels-coefficients can be expressed in terms of elastic and photoelastic constants as

$$p_{11} = \frac{2}{n_0^3}\frac{E}{(1+\nu)(1-2\nu)}\left[(1-\nu)C_1 + 2\nu C_2\right], \qquad (3.6)$$

$$p_{12} = \frac{2}{n_0^3}\frac{E}{(1+\nu)(1-2\nu)}(\nu C_1 + C_2). \qquad (3.7)$$

In table 3.1, the values of Young's modulus E, Poisson's ratio ν and the photoelastic coefficients C_1 and C_2 for fused silica at a wavelength of 546 nm are compiled according to Primak and Post [36]. The corresponding Pockel's coefficients can be calculated with equations (3.6) and (3.7) and the elastic and elastooptic parameters given in table 3.1. The result is $p_{11} = 0.121$ and $p_{12} = 0.271$, which is in reasonable agreement with the experimental values of $p_{11} = 0.126$ and $p_{12} = 0.26$ given by Borrelli et al. for bulk glasses at 632 nm [37] and $p_{11} = 0.113$ and $p_{12} = 0.252$ given by Bertholds et al. for fibers at 632 nm [38].

The dispersion of the Pockel's coefficients has been presented in the wavelength range from 457.9 to 676.4 nm by Biegelsen and Zesch [39] and is

constant	value	unit
Young's modulus E	0.76×10^{11}	N/m^2
Poisson's ratio ν	0.164	
Extraordinary photoelastic constant C_1	0.65×10^{-12}	m^2/N
Ordinary photoelastic constant C_2	4.22×10^{-12}	m^2/N

Table 3.1: *Elastic and photoelastic constants for vitreous silica at room temperature [36]. The photoelastic constants have been determined for a wavelength of 546 nm.*

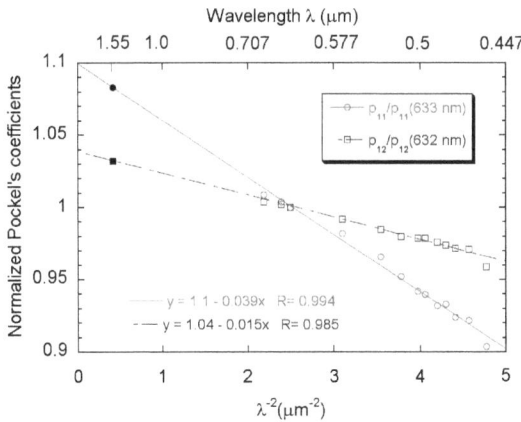

Figure 3.1: *Dispersion of the Pockel's coefficient according to [39]. The Pockel's coefficient at $\lambda = 1.55\,\mu m$ can be estimated by extrapolating the results in the wavelength range from 457.9 to 676.4 nm.*

sketched in Fig. 3.1. The values of the Pockel's coefficients have been normalized to their value at 633 nm. As a quadratic expansion in energy is expected well below the bandgap, the data depends linearly on λ^{-2} and can thus be extrapolated to obtain the Pockel's coefficients for a wavelength of 1.55 nm. The extrapolated values of $p_{11}/p_{11}(633\,\text{nm}) = 1.083$ and $p_{12}/p_{12}(633\,\text{nm}) = 1.032$ are depicted in black in Fig. 3.1.

If we assume $p_{11}(633\,\text{nm}) = 0.121$ and $p_{12}(633\,\text{nm}) = 0.271$, the Pockel's coefficient at 1.55 µm read $p_{11}(1550\,\text{nm}) = 0.131$ and $p_{12}(1550\,\text{nm}) = 0.279$. The corresponding photoelastic constants are found by solving equation (3.6) and (3.7) and yield $C_1 = 0.79 \times 10^{12}\,\text{m}^2/\text{N}$ and $C_2 = 4.19 \times 10^{12}\,\text{m}^2/\text{N}$, respectively. The relative photoelastic constant at 1.55 µm is $C = C_2 - C_1 = 3.4 \times 10^{12}\,\text{m}^2/\text{N}$, which is about 5% smaller than the value found for 548 nm from table 3.1. Namihira reports a decrease of the relative photoelastic constant by about 10% when the wavelength is increased from 0.632 µm to 1.55 µm [40]. Our estimate for the Pockel's coefficients at a wavelength of 1.55 µm thus agrees reasonably with experimental data.

3.3 Irradiation-induced index changes in optical fibers

3.3.1 Densification-induced refractive index changes

Using the elastooptic equations (3.3), the change in refractive index with volume can be expressed as

$$\Delta n = \frac{1}{3}\sum_i \Delta n_{ii} = -\frac{n_0^3}{2}\left(\frac{p_{11} + 2p_{12}}{3}\right)\frac{\Delta V}{V_0} \approx -0.341\frac{\Delta V}{V_0}, \quad (3.8)$$

where the approximation is valid for a wavelength of 1.55 µm. Equation (3.8) has been used by Borrelli et al. [41] as well as by Yablon et al. [42] for the calculation of refractive index response from inelastic density changes.

An alternative approach to express the refractive index dependence on density for a given material is the so-called Lorentz-Lorenz relation [43]

$$\frac{n_0^2 - 1}{n_0^2 + 2} = \frac{4\pi}{3}N\alpha = \frac{4\pi}{3}\frac{\alpha}{V_0}, \quad (3.9)$$

where α the mean polarizability, N the number of oscillators per unit volume, and V_0 the volume of a unit cell. To get information about the change in index with unit volume, equation (3.9) is differentiated with respect to V_0, resulting in

$$\Delta n = \frac{(n_0^2 - 1)(n_0^2 + 2)}{6\,n_0}(\Omega - 1)\frac{\Delta V}{V_0}, \quad (3.10)$$

where $\Omega = (\Delta\alpha/\alpha)/(\Delta V/V_0)$ is the change of polarizability with compaction.

When no change in polarizability with volume is assumed, i.e. $\Omega = 0$, equations (3.8) and (3.10) can directly be compared. For a wavelength of $1.55\,\mu$m, the proportional constant between index change and dilatation is -0.33 with equation (3.8) and -0.51 with equation (3.10). This discrepancy has already been reported by Pockels in 1902 [44] and can be explained with a reduction in polarizability with compression [36, 45]. By comparison of equation (3.8) with (3.10), a virtual value of Ω can be found for the description of index change with the elastooptic equations. For a wavelength of $1.55\,\mu$m, this virtual value yields $\Omega = 0.33$.

process	thermally	UV-irradiated	γ-irradiated	neutrons	Pockel's law
Ω	0.19	0.23	0.22	0.23	0.33
Ref.	[25]	[46]	[47]	[48]	Eq. (3.8)

Table 3.2: *Experimentally determined values for the change of polarizability with compaction Ω. In the rightmost column, a virtual Ω is presented that is found by comparison of equations (3.8) and (3.10) for a wavelength of $1.55\,\mu$m.*

In general, the change in polarizability with compaction is determined empirically by comparison of equation (3.10) with experimental data. In table 3.2, different values of Ω found for different compaction methods are compiled for comparison. The higher the value of Ω, the lower is the index response for a given amount of compaction. In Fig. 3.2, the change in index with density $\rho/\rho_0 = 1 - \Delta V/V$ according to the values of Ω presented in table 3.2 are illustrated graphically. As all experimental results have been derived for a wavelength of 632 nm, the theoretical index changes according to equations (3.8) and (3.10) are also calculated for this wavelength. All experiments result in index changes higher than predicted by the elastooptic relationship (3.8) and lower than given by the Lorentz-Lorenz equation (3.10) with $\Omega = 0$. For the experimentally obtained values of Ω, however, a non-compaction induced part, introduced i.e. by changes in the color center population, might also contribute to the total refractive index change. To exclude any influence of this type in the following, we use the most conservative expression (3.10) with $\Omega = 0.33$ to express the dependence of refractive index on compaction.

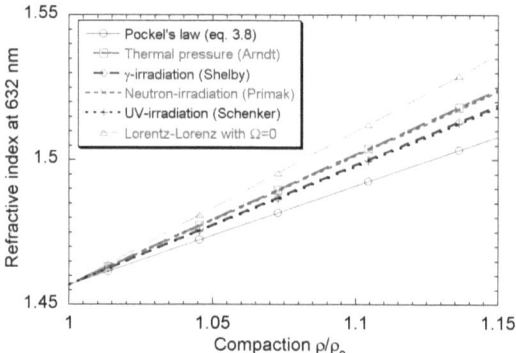

Figure 3.2: *Comparison of densification data for different experimental methods. For all methods, the increase in index change is higher than predicted by (3.8).*

3.3.2 Elastic response and net index change

When the core of the fiber compacts during irradiation, new elastic stresses arise in the core according to equation (2.40). These stresses, in turn, change the refractive index due to the photoelastic effect (3.5). The total index change thus is the superposition of a compaction-induced index change and an index change caused by the elastic response to this compaction. Using equation (3.10) for the inelastic contribution and equation (3.5) with equations (2.40), (2.22) and (2.23) for the elastic response, the overall or net transverse index change in the fiber core reads

$$\Delta n_\perp^{\text{tot}} = \Delta n^{\text{comp}} + \Delta n_\perp^{\text{el}} \tag{3.11}$$

$$= \left[\frac{(n_0^2 - 1)(n_0^2 + 2)}{6\, n_0} (\Omega - 1) + \frac{3C_2 + C_1}{6} \frac{E}{(1-\nu)} \right] \frac{\Delta V}{V} \tag{3.12}$$

$$= \left[\frac{(n_0^2 - 1)(n_0^2 + 2)}{6\, n_0} (\Omega - 1) \frac{3(1-\nu)}{E} - \frac{3C_2 + C_1}{2} \right] \Delta\sigma_{zz}, \tag{3.13}$$

where equation (2.40) has been used to convert compaction into axial core stress change. Here, we use the transverse index component $\Delta n_\perp^{\text{tot}}$, as the light propagating along the fiber axis is only influenced by this component of the refractive index tensor. In Fig. 3.3, the compaction-induced, photoelastic, and net-index change are illustrated as a function of compaction for $\Omega = 0.33$ and a wavelength of $1.55\,\mu\text{m}$. The photoelastic component com-

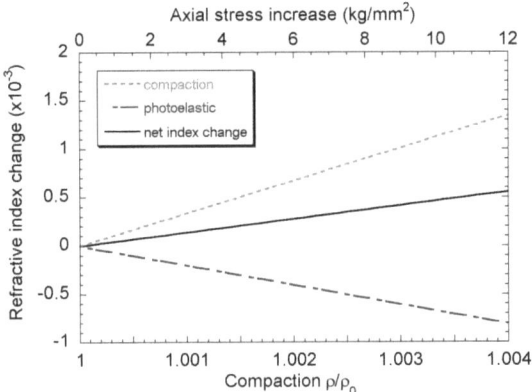

Figure 3.3: *Elastic and inelastic contribution to the total index change. The net index-change is reduced by almost 60% by the photoelastic response to compaction.*

pensates for almost 60% of the compaction-induced index change, which is in agreement with experimental results for germanium-doped fibers [5].

Using equation (3.13), a linear relation of $\Delta\sigma_{zz}/\Delta n_{\perp}^{tot} = 2.2 \times 10^4$ kg/mm^2 is found between the net compaction induced index change and the axial stress increase. This value can further be used to estimate the contribution of compaction to the overall irradiation-induced index change: if all index change was due to compaction, the ratio between stress and total index increase must be 2.2×10^4 kg/mm^2, whereas it must be smaller if a part of the index change was due to changes in the color center population.

Fonjallaz *et al.* determined a linear relationship of $0.9 \pm 0.1 \times 10^4$ kg/mm^2 between stress and index change [4], which is less than half of the theoretical value of 2.2×10^4 kg/mm^2 given by equation (3.12). However, as they correlated the stress change with the total index change (including the contribution of color centers) and not only with the compaction induced index change, this discrepancy does not contradict our result.

Poumellec *et al.* calculated the relationship between the mean axial stress change and the modulation index for a fiber Bragg grating to be 0.83×10^4 kg/mm^2 [21]. However, they used a value of $\Omega = 0$ in the Lorentz-Lorenz equation (3.10) and not the more conservative estimate of $\Omega = 0.33$ used within this work. If we also include $\Omega = 0$ in equation (3.10), we get a value of 0.98×10^4 kg/mm^2, which agrees reasonably well with Poumellec's result.

3.4 Measurement of photosensitivity

Since the discovery of fiber Bragg gratings, a number of techniques have been developed to determine the UV-induced refractive index change. Due to the small amount of index changes especially at the beginning of the writing process, conventional methods used to determine the index profile of optical fibers, as refractive near field (RNF) measurements or transverse interferometric methods (TIM), are inappropriate [49]. Moreover, they do not allow to follow the index changes online during the irradiation process.

In contrast, the observation of the spectral response of a grating during its inscription allows principally the online-determination of both, the gratings average index and index modulation change [50]. As this method is simple, reliable and widely established for the characterization of photosensitivity, it was also mostly used within this work. Alternative methods using in-fiber interferometric configurations [51, 52] allow the determination of index changes for purely bulk exposure.

A complementary method to obtain information about refractive index profiles and their light-induced changes is the profiling of the cleaved and subsequently etched fiber endface using an Atomic Force Microscope (AFM). The outstanding advantage of this measurement technology is the spatial resolution of the captured profiles, which is in the range of several tenth of nanometers. For germanium-doped fibers it has been shown that the etch profiles can be directly related to the core-cladding refractive index profile [53, 54]. However, to get information about the absolute quantitative values of refractive index, calibration measurements using conventional optical methods are necessary [53].

3.4.1 Bragg grating inscription

Phase mask technique

A variety of methods have been reported for the fabrication of fiber Bragg gratings, that can be principally divided in interferometric and phase mask based methods [50]. The former methods generally allow a more flexible adjustment of the grating characteristics, whereas the latter are more stable. In this work, a phase mask technique was used for Bragg grating inscription. The configuration of the setup is illustrated in Fig. 3.4. The phase mask diffracts the light of the incoming UV-beam. For most phasemasks, the diffraction in the ± first orders is maximized. The interference of the two orders results in a periodic intensity modulation in the region directly behind the phase mask. The size and contrast of the region of interference depends strongly on the temporal and spatial coherence of the illuminating laser source. The period of the interference pattern is half the period of the

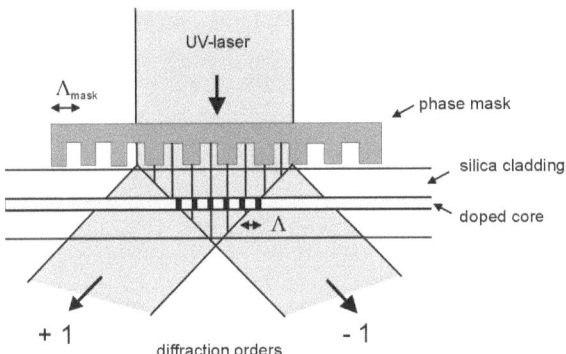

Figure 3.4: Bragg grating inscription using the phase mask technique. The fiber core is placed into the interference pattern of the light diffracted by the phase mask.

phase mask. When a fiber is placed in the interference region, the refractive core index of the regions at the interference maxima increases considerably stronger than those at the minima, and thus, a grating having the same period as the interference pattern is generated inside the fiber core. The grating is in general assumed to be sinusoidal in shape and is characterized by the refractive index distribution

$$\Delta n(z) = \Delta n_{\text{dc}}(z) + \Delta n_{\text{ac}}(z) \cos\left(\frac{2\pi}{\Lambda} z\right), \quad (3.14)$$

where Λ is the grating period, Δn_{dc} the average refractive index change and Δn_{ac} the index modulation amplitude. The refractive index distribution can principally be derived from the spectrum of the grating.

Bragg grating spectra

The wavelength response of optical waveguides to perturbations of the refractive index profile can in general be described by a coupled mode theory [55, 56]. For periodic perturbations of the core index profile, the maximum reflectivity occurs at the Bragg wavelength

$$\lambda_B = 2 n_{\text{eff}} \Lambda, \quad (3.15)$$

where n_{eff} is the effective core index and can be approximated by [57]

$$n_{\text{eff}}^2 = n_{Cl}^2 + \left(\frac{\lambda}{2\pi \varrho}\right)^2 (1.1428\, V - 0.996)^2. \quad (3.16)$$

51

In (3.16), n_{Cl} is the cladding refractive index, ϱ the core radius and

$$V = \frac{2\pi}{\lambda}\varrho\sqrt{n_{Co}^2 - n_{Cl}^2} \qquad (3.17)$$

the normalized frequency of the fiber with n_{Co} the core index before irradiation. The error introduced by equation (3.16) is smaller than 0.2% for $1.5 \leq V \leq 2.5$ [57].

For uniform gratings, where both Δn_{dc} and Δn_{ac} do not depend on the axial position, the maximum reflectivity of the grating is given by [56]

$$R_{max} = \tanh^2(\eta\frac{\pi\Delta n_{ac}}{\lambda}L), \qquad (3.18)$$

where L is the grating length and η the overlap-factor, characterizing the core power confinement of the fundamental mode. The overlap factor can be determined from the mode field diameter MFD according to [57]

$$\eta = 1 - \exp\left[-2\left(\frac{\varrho}{\text{MFD}}\right)^2\right], \qquad (3.19)$$

where the MFD can be expressed in terms of the normalized frequency (3.17) as [57]

$$\frac{MFD}{\varrho} = 0.65 + 1.619\,V^{-1.5} + 2.879\,V^{-6}. \qquad (3.20)$$

For standard telecommunication fibers, the overlap-factor is about $\eta = 0.83$.

Equations (3.15) and (3.18) allow the evaluation of the UV-induced index changes according to

$$\Delta n_{dc} = \frac{n_{\text{eff}}}{\eta}\frac{\Delta\lambda_B}{\lambda_B}, \qquad (3.21)$$

$$\Delta n_{ac} = \frac{\lambda_B}{\eta\pi L}\text{atanh}(\sqrt{R_{max}}), \qquad (3.22)$$

where λ_B is the wavelength where the first appearance of the Bragg peak is observed right at the beginning of the irradiation. In Fig. 3.5, transmission spectra recorded during the Bragg grating inscription process are illustrated. The maximum reflectivity is related to the minimum transmittivity by $T_{min} = 1 - R_{max}$, when losses at the reflection peak are neglegible. The shift in Bragg wavelength $\Delta\lambda_B$ is shown for two different irradiation times in Fig. 3.5. The longer the irradiation time, the bigger the shift in Bragg wavelength, and the stronger the strength in rejection. For the determination of refractive index changes, the length of the grating should be chosen appropriately to exclude saturation effects. For strong Bragg gratings with reflectivities $\geq 90\%$, even large changes in index have only a small effect on the peak depth.

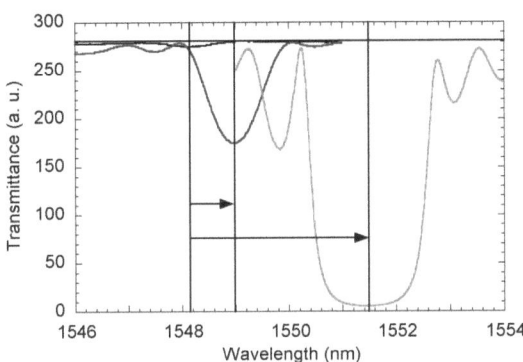

Figure 3.5: *Bragg wavelength shift and transmittivity changes during the inscription of a Bragg grating. The observed spectral changes allow the determination of refractive index changes according to equations (3.21) and (3.22).*

3.4.2 Atomic Force Microscopy

State of the art

Atomic force microscopy (AFM) of chemically etched fiber end-faces yields topographic information on a nanometer scale due to differential chemical etching of doped and un-doped silica. For germanium-doped fibers it has been shown that the etch rate can be related to the dopant concentration profile as well as to the core-cladding refractive index profile [53, 54]. On a large scale, the etch depth is independent of processing conditions, e.g. fiber drawing tension [54]. However, different ellipticities between the etch profile of an elliptical core fiber and its corresponding preform index-profile were observed in [58, 59]. The exact dependence of etch rate on chemical composition (core dopants) on the one hand and drawing-induced properties of the fiber core on the other hand thus demands further clarification.

In addition, the origins of etch rate changes due to UV-illumination [60, 61] still require an appropriate explanation. Inniss *et al.* reported an asymmetric etch profile for hydrogen-loaded standard telecommunication fiber caused by side exposure to UV at 244 nm [60]. For hydrogenated phosphorus-doped fibers, preferential etching at the core-cladding boundary has been observed by Canning *et al.* and attributed to stress-assisted UV-initiated bond breaking at the interface [61]. However, no systematic study on etch rate changes with laser-irradiation have been presented so far.

Measurement procedure

To prepare the fibers for AFM-measurements, the coating is removed and the fibers are cleaved with a Fujikura CT-07BS high precision fiber cleaver. The quality of the obtained fiber endfaces is controlled with a conventional optical microscope. In the next step, up to six fibers are fixed magnetically in six V-grooves of a fiber holder. The holder is put on top of a tub containing the hydrofluoric acid, so that the ends of the six fibers are dipped in the acid. The concentration of the hydrofluoric acis as well as the etch time should be chosen in accordance with the dopant of the fiber core and its concentration. For example, we found that germanium- and phosphorus-doped cores etch considerably faster than nitrogen-doped fiber cores. After the etch, the maximum height difference between core and cladding should not exceed about 1000 nm. For higher height differences, the tip of the AFM cantilever does not reach down to the bottom of the hole, as the sides of the V-shaped cantilever already hit the sample before. When the etch time is over, the fibers are rinsed in distilled water and isopropanol to immediately stop the etching process.

To mount the samples for the AFM-measurements, they are glued on an aluminum bar and vertically aligned. The viscosity of the glue should be high enough to avoid that it soars up along the fiber, which might at the end even lead to a contamination of the cleaved and etched fiber endface. The AFM (TopoMetrix Explorer) is operated in contact mode in air with a V-shaped nitride cantilever. A typical example for a thus obtained profile is illustrated in Fig. 3.6. The doped core layers are etched faster than the surrounding pure silica inner cladding layers. The dip in the center of the core is caused by dopant out diffusion during collapsing of the preform tube.

3.5 Summary and Conclusions

The origins of photosensitivity have not been clarified in detail yet and are a matter of ongoing scientific discussion. However, two models have been proved to account significantly to the observed index change: the color-center model and the compaction model. Within this work, the amount of compaction of the fiber core is determined by measuring compaction-induced axial stress changes. The refractive index dependence on compaction is estimated by comparison with compaction data found with different densification methods. The overall index change does not only depend on the amount of compaction, but also on the elastic response of the stress profile to compaction and its contribution to the index via the photoelastic effect. Both index changes and stress changes are linearly related to the densification.

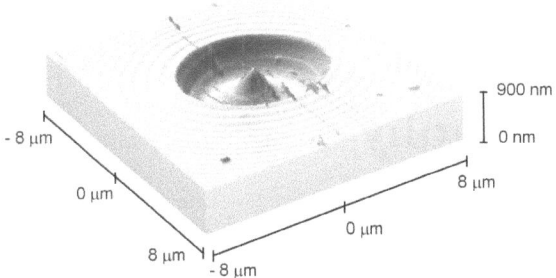

Figure 3.6: *AFM image of a phosphorus-doped fiber after etching in hydrofluoric acid for* 180 *seconds. The core is observed as crater since it is etched faster than the concentric inner cladding SiO_2 layers*

The elastic response to the core glass compaction contributes negatively to the overall index change.

To compare the amount of compaction to refractive index changes, the change in index must be measured. Two different methods have been used within this work for this purpose: Bragg grating inscription with simultaneous measurement of the spectral response and AFM profiling of etched fiber end-faces. The first method gives quantitative results of the index level, whereas the second method provides information about light-induced etch rate changes with high spatial resolution.

Bibliography

[1] K. O. Hill, Y. Fujii, D. C. Johnson, and B. S. Kawasaki. Photosensitivity in optical fiber waveguides: Application to reflection filter fabrication. *Applied Physics*, 32(10):647–649, 1978.

[2] V. B. Neustruev. Colour centres in germanosilicate glass and optical fibres. *Journal of Physics: Condensed Matter*, 6:6901–6936, 1994.

[3] J. Nishii. Permanent index changes in Ge-SiO$_2$ glasses by excimer laser irradiation. *Materials Science & Engineering B*, 54:1–10, 1998.

[4] P. Y. Fonjallaz, H. G. Limberger, R. P. Salathe, F. Cochet, and B. Leuenberger. Tension increase correlated to refractive-index change in fibers containing UV-written Bragg gratings. *Optics Letters*, 20(11):1346–1348, 1995.

[5] H. G. Limberger, P. Y. Fonjallaz, R. P. Salathe, and F. Cochet. Compaction- and photoelastic-induced index changes in fiber Bragg gratings. *Applied Physics Letters*, 68(22):3069–3071, 1996.

[6] P. K. Bachmann, W. Hermann, H. Wehr, and D. U. Wiechert. Stress in optical waveguides. II. Fibers. *Applied Optics*, 26(7):1175–1182, 1987.

[7] Y. Hibino and H. Hanafusa. Formation mechanism of defect centers in GeO$_2$-doped silica glass. *Journal of Non-Crystalline Solids*, 95–96(1):343–350, 1987.

[8] P. J. Lemaire, R. M. Atkins, V. Mizrahi, and W. A. Reed. High pressure H$_2$ loading as a technique for achieving ultrahigh UV photosensitivity and thermal sensitivity in GeO$_2$ doped optical fibres. *Electronics Letters*, 29(13):1191–1193, 1993.

[9] P. Niay, P. Bernage, S. Legoubin, M. Douay, W. X. Xie, J. F. Bayon, T. Georges, M. Monerie, and B. Poumellec. Behaviour of spectral transmissions of Bragg gratings written in germania-doped fibres: writing and erasing experiments using pulsed or cw uv exposure. *Optics Communications*, 113(1–3):176–192, 1994.

[10] D. L. Griscom. Optical properties and structure of defects in silica glass. *Journal of the Ceramic Society of Japan*, 99:899–916, 1991.

[11] G. R. Atkins, W. Zhen Hua, D. R. McKenzie, M. G. Sceats, S. B. Poole, and H. W. Simmons. Control of defects in optical fibers-a study using cathodoluminescence spectroscopy. *Journal of Lightwave Technology*, 11(11):1793–1801, 1993.

[12] M. Gallagher and U. Osterberg. Spectroscopy of defects in germanium-doped silica glass. *Journal of Applied Physics*, 74(4):2771–2778, 1993.

[13] L. Dong, J. Pinkstone, P. S. J. Russell, and D. N. Payne. Ultraviolet absorption in modified chemical vapor deposition preforms. *Journal of the Optical Society of America B*, 11(10):2106–2111, 1994.

[14] P. Russell, D. P. Hand, Y. T. Chow, and L. J. Poyntz Wright. Optically-induced creation, transformation and organisation of defects and colour-centres in optical fibres. *Proceedings of the SPIE*, 1516:47–54, 1991.

[15] D. L. Williams, S. T. Davey, R. Kashyap, J. R. Armitage, and B. J. Ainslie. Direct observation of UV induced bleaching of 240 nm absorption band in photosensitive germanosilicate glass fibres. *Electronics Letters*, 28(4):369–371, 1992.

[16] L. Dong, J. L. Archambault, L. Reekie, P. St. J. Russell, and D. N. Payne. Photoinduced absorption change in germanosilicate preforms: evidence for the color-center model of photosensitivity. *Applied Optics*, 34(18):3436–3440, 1995.

[17] M. J. F. Digonnet. A Kramers-Kronig analysis of the absorption change in fiber gratings. *Proceedings of the SPIE*, 2841:109–120, 1996.

[18] M. G. Sceats, G. R. Atkins, and S. B. Poole. Photolytic index changes in optical fibers. *Annual Review Material Science*, 23:381–410, 1993.

[19] J. P. Bernardin and N. M. Lawandy. Dynamics of the formation of Bragg gratings in germanosilicate optical fibers. *Optics Communications*, 79(3–4):194–199, 1990.

[20] B. Poumellec, P. Guenot, I. Riant, P. Sansonetti, P. Niay, P. Bernage, and J. F. Bayon. UV induced densification during Bragg grating inscription in Ge:SiO_2 preforms. *Optical Materials*, 4(4):441–449, 1995.

[21] B. Poumellec, P. Niay, M. Douay, and J. F. Bayon. The UV-induced refractive index grating in Ge:SiO_2 preforms: Additional CW experiments

and the macroscopic origin of the change in index. *Journal of Physics D (Applied Physics)*, 29(7):1842–1856, 1996.

[22] D. Wong, S. B. Poole, and M. G. Sceats. Stress relief: proof of the mechanism of photo-induced index change. *Integrated Photonics Research*, pd16:408–412, 1992.

[23] C. Z. Tan and J. Arndt. The refractive index of silica glass and its dependence on pressure, temperature, and the wavelength of the incident light. In H. S. Nalwa, editor, *Silicon based materials and devices*. Academic Press, 2001.

[24] N. Kitamura, I. Y. Toguchi, S. Funo, H. Yamashita, and M. Kinoshita. Refractive index of densified silica glass. *Journal of Non-Crystalline Solids*, 159(3):241–245, 1993.

[25] C. Z. Tan, J. Arndt, and H. S. Xie. Optical properties of densified silica glass. *Physica B*, 252:28–33, 1998.

[26] E. M. Dianov, V. M. Mashinsky, V. B. Neustruev, O. D. Sazhin, V. V. Brazhkin, and V. A. Sidorov. Optical absorption and luminescence of germanium oxygen-deficient centers in densified germanosilicate glass. *Optics Letters*, 22(14):1089–1091, 1997.

[27] D. Brewster. On the effects of simple pressure in producing that species of crystallisation which forms two oppositely polarized images. *Philosophical transactions of the Royal Society of London*, 105:60–65, 1815.

[28] F. E. Neumann. Die Gesetze der Doppelbrechung des Lichts in comprimierten oder ungleichförmig unkrystallinischen Körpern. *Abhandlungen der Königlichen Academie der Wissenschaften zu Berlin Part II*, pages 1–254, 1841.

[29] J. C. Maxwell. On the equilibrium of elastic solids. *Transactions of the Royal Society of Edinburgh*, 20:87–120, 1853.

[30] E. G. Coker and L. N. G. Filon. *A Treatise on Photoelasticity*. Camebridge University Press, 1931.

[31] S. E. A. Bayoumi and E. K. Frankl. Fundamental relations in photoplasticity. *British Journal of Applied Physics*, 4(10):306–310, 1953.

[32] J. Javornicky. *Photoplasticity*. Elsevier Scientific Publishing Company, 1974.

[33] F. Pockels. *Lehrbuch der Kristalloptik*. B. G. Teubner, 1906.

[34] S. P. Timoshenko and J. N. Goodier. *Theory of Elasticity*. McGraw-Hill, 1970.

[35] A. C. Ugural and S. K. Fenster. *Advanced strength and Applied Elasticity*. Elsevier, 1987.

[36] W. Primak and D. Post. Photoelastic constants of vitreous silica and its elastic coefficient of refractive index. *Journal of Applied Physics*, 30(5):779–788, 1959.

[37] N. F. Borrelli and R. A. Miller. Determination of the individual strain-optic coefficients of glass by an ultrasonic technique. *Applied Optics*, 7(5):745–750, 1968.

[38] A. Bertholds and R. Dändliker. Determination of the individual strain-optic coefficients in single-mode optical fibers. *Journal of Lightwave Technology*, 6(1):17–20, 1988.

[39] D. K. Biegelsen and J. C. Zesch. Optical frequency dependence of the photoelastic coefficients of fused silica. *Journal of Applied Physics*, 47(9):4024–4025, 1976.

[40] Y. Namihira. Opto-elastic constant in single mode optical fibers. *Journal of Lightwave Technology*, 3(5):1078–1083, 1985.

[41] N. F. Borrelli, D. C. Allan, and R. A. Modavis. Direct measurement of 248- and 193-nm ecximer-induced densification in silica-germania waveguide blanks. *Journal of the Optical Society of America B*, 16(10):1672–1679, 1999.

[42] A. D. Yablon. Optical and mechanical effects of frozen-in stresses and strains in optical fibers. *IEEE Journal of Selected Topics in Quantum Electronics*, 10(2):300–311, 2004.

[43] C. Z. Tan and J. Arndt. The mean polarizability and density of glasses. *Physica B*, 229:217–224, 1997.

[44] F. Pockels. Über die Änderung des optischen Verhaltens verschiedener Gläser durch elastische Deformation. *Annalen der Physik*, 7:745–771, 1902.

[45] H. Mueller. Theory of the photoelastic effect of cubic crystals. *Physical Review*, 47:947–957, 1935.

[46] R. E. Schenker and W. G. Oldham. Ultraviolet-induced densification in fused silica. *Journal of Applied Physics*, 82(3):1065–1071, 1997.

[47] J. E. Shelby. Effect of radiation on the physical properties of borosilicate glasses. *Journal of Applied Physics*, 51(5):2561–2565, 1980.

[48] W. Primak. Fast-neutron-induced changes in quartz and vitreous silica. *Journal of Applied Physics*, 110(6):1240–1254, 1958.

[49] K. W. Raine, J. G. N. Baines, and D. E. Putland. Refractive index profiling-state of the art. *Journal of Lightwave Technology*, 7(8):1162–1169, 1989.

[50] Raman Kashyap. *Fiber Bragg Gratings*. Academic Press, 1999.

[51] E. M. Dianov, S. A. Vasiliev, A. S. Kurkov, O. I. Medvedkov, and V. N. Protopopov. In-fiber Mach-Zehnder interferometer based on a pair of long-period gratings. In *Proc. European Conf. on Optical Communication, ECOC'96*, 1996. 65-69 vol.1.

[52] D. Johlen, H. Renner, A. Ewald, and E. Brinkmeyer. Fiber Bragg grating Fabry-Perot interferometer for a precise measurement of the UV-induced index change. In *Proc. European Conf. on Optical Communication, ECOC'98*, 1998. 393-394 vol.1.

[53] P. Pace, S. T. Huntington, K. Lyytikainen, A. Roberts, and J. D. Love. Refractive index profiles of Ge-doped optical fibers with nanometer spatial resolution using atomic force microscopy. *Optics Express*, 12(7):1452–1457, 2004.

[54] Q. Zhong and D. Inniss. Characterization of the lightguiding structure of optical fibers by atomic force microscopy. *Journal of Lightwave Technology*, 12(9):1517–1523, 1994.

[55] Dietrich Marcuse. *Theory of Dielectric Optical Waveguides*. Academic Press, INC., second edition, 1991.

[56] T. Erdogan. Fiber grating spectra. *Journal of Lightwave Technology*, 15(8):1277–1294, 1997.

[57] L. B. Jeunhomme. *Single-mode fiber optics: principles and applications*. Marcel Dekker, Inc., 1990.

[58] S. T. Huntington, P. Mulvaney, A. Roberts, K. A. Nugent, and M. Bazylenko. Atomic force microscopy for the determination of refractive index profiles of optical fibers and waveguides: a quantitative study. *Journal of Applied Physics*, 82(6):2730–2734, 1997.

[59] S. T. Huntington. *Advances in Photonics Device Characterization using Scanning Probe Microscopy.* PhD thesis, School of Physics, University of Melbourne, 1998.

[60] D. Inniss, Q. Zhong, A. M. Vengsarkar, W. A. Reed, S. G. Kosinski, and P. J. Lemaire. Atomic force microscopy study of UV-induced anisotropy in hydrogen-loaded germanosilicate fibers. *Applied Physics Letters*, 65(12):1528–1530, 1994.

[61] J. Canning, K. Sommer, M. Englund, and S. Huntington. Direct evidence of two types of UV-induced glass changes in silicate based optical fibers. *Advanced Materials*, 13(12-13):970–973, 2001.

Chapter 4

Stress and strain profile determination in optical fibers

One of the main objectives of this thesis was the planning, realization and characterization of a setup capable to determine residual stresses and strains in optical fibers. In the first section of this chapter, we give a brief summary about the history and state of the art of such measurement techniques in optical fibers. We then sketch the theoretical background necessary for a proper understanding of the measurement principle in the second section. We show that the determination of the complete strain tensor can be achieved by integrating the fibers birefringence over its cross-section. Furthermore, we interpret the measured birefringence in terms of both elastic and inelastic strain and not, as in hitherto existing setups, only in terms of elastic strain or stress, respectively. When the inelastic strain is assumed to be constant over the complete cross section, one can accurately distinguish between the elastic and inelastic strain contribution. From the elastic strain contribution, the corresponding stress profiles can subsequently be calculated. As the basis for all calculations is the fibers transverse birefringence profile, we present in the third section a polariscope which captures birefringence data with both high sensitivity and high spatial resolution. The calibration of the setup is checked by straining a fiber and measuring the corresponding stress changes. The section terminates with the assessment of the setups reproducibility and spatial resolution.

4.1 Review of existing measurement techniques

The application of polarimetric methods for the determination of quenched stresses in glass cylinders and fibers has been reported for the first time about fifty years ago [1, 2]. The first measurements on optical fibers suffered from the low spatial resolution as well as from a limited precision in the determination of the integrated birefringence [3, 4]. Saunders [5] and Nagano et al. [6] measured stress and birefringence changes in optical fibers due to external forces applied to the fiber. Shibata et al. [7] reported about the measurement of residual stress-induced retardation profiles in optical preforms. They used a Sénarmont-compensator [8] to determine the stress-induced change in polarization with high precision. The same measurement principle was adopted by Lagakos et al. [9] to optical fibers. They also presented residual stress profiles calculated from the retardation data, but as they ignored the radial stress dependence, their stress results were erroneous. The correct link between stress and retardation data was given by Chu and Whitbread [10, 11] in 1982. They showed that the axial stress profile can directly be obtained from the Abel transform of the derivative of the retardation. The use of the Abel-transform had also been suggested by Shibata et al. [7] and Anderssen et al. [12] to determine the thermal expansion coefficient distribution from the retardation data of optical preforms. In the following years, several groups presented setups similar to the one of Chu and Whitbread for both preform and fiber stress profiling [13–18].

In 1985, Abe et al. demonstrated the measurement of two-dimensional axial stress profiles in optical fibers [19, 20]. They actually used the same setup as Chu and Whitbread, but recorded the retardation data of the fiber for a number of different projection angles by rotating the fiber over a range of 180°. A filtered backprojection method, also known as inverse Radon-transform, was used to obtain two-dimensional axial stress profiles from the retardation data. The same method was adopted by Puro and Kell to determine the axial stress profile in optical fiber preforms [21]. They further presented an algorithm that allowed the assessment of the complete stress tensor from the axial stress profile and has already been presented in subsection 2.1.3. Tomographic measurements have also been applied to optical fibers by Park et al. [22, 23]. For data acquisition, they used a CCD camera and incoherent illumination, which made their setup considerably faster in comparison to earlier systems using a scanning technique. In addition, they used a fixed analyzer and a rotating polarizer, a setup which is basically equivalent to earlier setups where rotating analyzers and fixed polarizers have been used. The advantage of the different architecture is the possibility to locate the polarizer directly next to the optical fiber, which results in a reduction of wavefront errors.

Within this work, a stress measurement setup was planned and realized, which is basically similar to the setup presented by Park et al.. The data is acquired in parallel using a CCD camera. A rotating polarizer is used to control the polarization of the illuminating light. However, in contrast to Park's setup, the analyzer is placed behind the imaging microscope objective, which made the objective's working distance a less critical parameter. In this configuration, the polarization is not allowed to be changed by the microscope objective, so special stress-free microscope objectives used for polarization contrast microscopy are used. To enable the tomographic measurement of stress, a fiber rotator with a rotation range of 240° is integrated in the setup.

So far, all groups concerned in the field have interpreted the birefringence data to be solely stress or, respectively, elastic strain induced. Here, we present for the first time an interpretation of birefringence data in terms of both elastic and inelastic strain.

4.2 Theory

4.2.1 Integrated phase retardation

When an optical ray passes through an anisotropic medium, its polarization is generally changed. The two components of the (transverse) electric field propagate with different velocities, as they experience different refractive indices [8]. In the following, we consider the light to propagate in positive x-direction and the main axes of birefringence to coincide with the axes of the coordinate system. The two components of the light vector then accumulate a phase retardation $\Re$ according to

$$d\Re(y) = dx\,[n_{zz}(x,y) - n_{yy}(x,y)] = dx\,B(x,y), \tag{4.1}$$

where we introduced the local birefringence B. Equation (4.1) is in the following applied to the geometry of an optical fiber as depicted in Fig. 4.1. Due to the circular shape of the fiber, the integration along the x-direction results in a retardation profile given by

$$\Re^{\mathrm{tot}}(y) = \Re^{\mathrm{el}}(y) + \Re^{\mathrm{in}}(y) = \int_{-\sqrt{R^2-y^2}}^{\sqrt{R^2-y^2}} \left[B^{\mathrm{el}}(x,y) + B^{\mathrm{in}}(x,y)\right]dx. \tag{4.2}$$

In equation (4.2) as far as in the following, we distinguish between elastic strain- respectively stress-induced birefringence B^{el} on the one hand and birefringence caused by inelastic strain B^{in} on the other hand. Their appropriate contributions $\Re^{\mathrm{el}}$ and $\Re^{\mathrm{in}}$ to the overall retardation $\Re^{\mathrm{tot}}$ are derived in more detail in the following.

Stress-induced retardation

Refractive index changes as a consequence of stress changes are described by equation (3.5). The elastic contribution to the total retardation is directly obtained by inserting equation (3.5) into equation (4.2)

$$\Re^{\mathrm{el}}(y) = C \int_{-\sqrt{R^2-y^2}}^{\sqrt{R^2-y^2}} (\sigma_{zz} - \sigma_{yy})\,dx = C \int_{-\sqrt{R^2-y^2}}^{\sqrt{R^2-y^2}} \sigma_{zz}\,dx, \tag{4.3}$$

where the photoelastic constant or Brewster-coefficient $C = C_2 - C_1$ has already been introduced. When the residual stresses are independent of the axial coordinate and no external stress is applied to the fiber, the integral over the y-component of the stress-tensor vanishes for both symmetric [24, 25] and

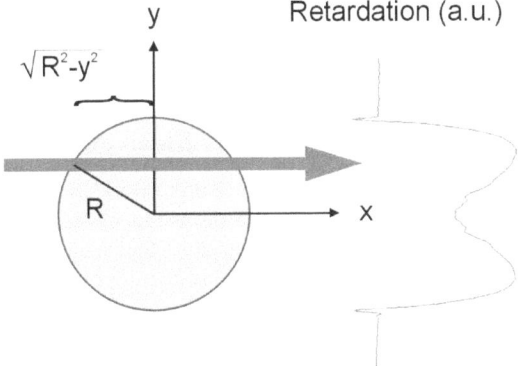

Figure 4.1: *The retardation profile of the fiber corresponds to the integrated birefringence of the fiber along the irradiating laser beam.*

asymmetric [21] cylinders. Equation (4.3) thus gives a direct link between the axial stress profile of an optical fiber and its corresponding projection data for both cylindrical symmetric and asymmetric fibers.

Inelastic strain-induced retardation

According to the model presented in subsection 2.2.2 of the second chapter, an inelastic strain anisotropy $\Delta\epsilon = \epsilon_{zz} - \epsilon_{\perp}$ is frozen into the fiber during the drawing process. The anisotropy is basically influenced by the material with the higher viscosity, which in general is the cladding for silicate fibers. As most of the fiber cross-sectional area is composed of the cladding, we assume the anisotropy to be constant over the fiber cross-section. The corresponding retardation experienced by the laser beam is found with equation (3.3) and yields

$$\mathfrak{R}^{\text{in}}(y) = \frac{n_0^3}{2} (p_{12} - p_{11}) \int_{-\sqrt{R^2-y^2}}^{\sqrt{R^2-y^2}} \Delta\epsilon \, dx \qquad (4.4)$$

$$= n_0^3 (p_{12} - p_{11}) \Delta\epsilon \sqrt{R^2 - y^2} \qquad (4.5)$$

The inelastic strain-induced retardation profile is thus semicircular, as evident from equation (4.5), and the diameter of the semicircle scales linearly with the inelastic strain anistropy. As we show below, the constant value of the inelastic strain anisotropy and the semicircular shape of the corresponding contribution to the overall retardation profile allows to discriminate between the inelastic and the elastic contribution.

4.2.2 Axial symmetric stress profiles: Abel-transform

For one-dimensional tomography of circular symmetric fibers, Abel-inversion, onion-peeling and filtered backprojection methods can most generally be used to recover radial fields from projection data [26]. The Abel-transform, however, does not only allow to give a closed analytical expression for the deconvolution process, but was also found to be superior in terms of ease of calculation, robustness and noise [26]. The problem of tomography of cylindrically symmetric objects has already analytically been solved early in the nineteenth century by the Norwegian mathematician Niels Henrik Abel [27]. Here, we use an alternative form of the Abel-transform,

$$\int_r^\infty F(\rho) \frac{\rho\, d\rho}{\sqrt{\rho^2 - r^2}} = G(r) \tag{4.6}$$

$$-\frac{2}{\pi} \int_\rho^\infty \frac{dG}{dr} \frac{dr}{\sqrt{r^2 - \rho^2}} = F(\rho) \tag{4.7}$$

that has been derived in more detail by Marcuse in [28]. $G(r)$ represents the projection data, whereas $F(\rho)$ is the field that shall be determined. To apply the Abel-transform pair (4.6) and (4.7) to our problem, we rewrite equations (4.3) and (4.5) in a cylindical coordinate system:

$$\mathfrak{R}^{\mathrm{el}}(y) = 2\,C \int_y^R \frac{\sigma_{zz}\, r}{\sqrt{r^2 - y^2}}\, dr \tag{4.8}$$

$$\mathfrak{R}^{\mathrm{in}}(y) = n_0^3\, (p_{12} - p_{11})\, \Delta\epsilon \int_y^R \frac{r}{\sqrt{r^2 - y^2}}\, dr. \tag{4.9}$$

Equations (4.6) and (4.7) allow us to assign $F(\rho)$ to σ_{zz} for the elastic case (4.8) and to 1 for the inelastic case (4.9). Due to the linearity of the Abel-transform, its application to the total retardation profile (4.2) yields

$$\sigma_{zz} + \frac{E}{1+\nu}\Delta\epsilon = -\frac{1}{\pi C} \int_r^R \frac{d\mathfrak{R}^{\mathrm{tot}}/dy}{\sqrt{y^2 - r^2}}\, dy, \tag{4.10}$$

where we made use of the relations (3.6) and (3.7) between the Pockel's coefficients p_{ij} and the photoelastic constants C_i. The Abel-inversion of the total retardation profile is thus the sum of the elastic stress profile and a constant offset proportional to the inelastic strain. This offset has been labeled by the oxymoron "inelastic stress" by Park et al. [29]. As the inelastic strain anisotropy $\Delta\epsilon$ can in principal be converted to a virtual axial stress by inserting equations (2.48) and (2.49) in (2.50), we adopt their interpretation in general and take the Abel-transform of the total retardation profile as the superposition of a real "elastic" and a virtual "inelastic" stress:

$$\tilde{\sigma}_{zz}^{\text{tot}} = \sigma_{zz}^{\text{el}} + \tilde{\sigma}_{zz}^{\text{in}}, \tag{4.11}$$

where the virtual "inelastic stress" is defined as

$$\tilde{\sigma}_{zz}^{\text{in}} = \frac{E}{1+\nu}\Delta\epsilon. \tag{4.12}$$

The tilde in equations (4.11) and (4.12) indicate that the stresses are only virtual, in analogy to the meaning of the quotation marks in the text.

The Abel-inversion of the total retardation profile gives the superposition of a real "elastic" stress and a virtual "inelastic" stress. For proper interpretation of the data, however, one should be capable to distinguish between the two contributions. Therefore, we apply St. Vernant's principle (2.16), which only holds for the elastic contribution, to equation (4.11). Thus, the elastic contribution in equation (4.11) vanishes and we find the expression

$$\tilde{\sigma}_{zz}^{\text{in}} = \frac{\int_A \tilde{\sigma}_{zz}^{\text{tot}} dA}{\int_A dA} \tag{4.13}$$

for the virtual "inelastic stress". The corresponding ("real") inelastic strain anisotropy can subsequently be deduced from equation (4.12). The inelastic strain anisotropy is therefore proportional to the mean value of the virtual total stress.

We demonstrate the measurement principle in the following for an SMF-28® standard telecommunication fiber. The retardation data of the fiber is shown on the top of Fig. 4.2. The maximum retardation in the cladding region is about 2 nm or 1.14°. The corresponding total stress profile $\tilde{\sigma}_{zz}^{\text{tot}}(r)$ is calculated with the Abel-inversion according to (4.10) and is depicted in the middle of the figure. Clearly, St. Vernant's principle (2.16) is violated for the overall stress profile, which indicates a significant amount of inelastic strain in the fiber. The virtual "inelastic stress", i.e. the mean value of the total stress is found with equation (4.13) to be $0.4\,\text{kg/mm}^2$. The corresponding strain anisotropy is $\Delta\epsilon = 4.4 \times 10^{-5}$. Both the stress profile and the inelastic strain anisotropy are depicted in the bottom of Fig. 4.2.

Once the stress profile has been separated from the virtual "inelastic stress", it can be inserted in equations (2.13) and (2.15) to determine the

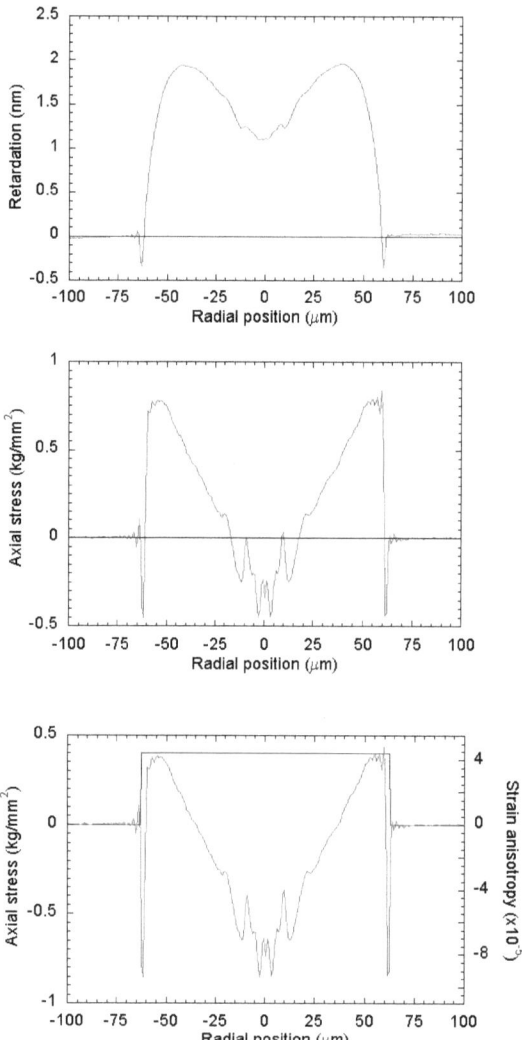

Figure 4.2: *Retardation profile obtained for an SMF-28® standard telecommunication fiber (above). The Abel-inversion of the retardation data gives the superposition of the stress profile with an constant virtual "inelastic stress" (middle). The two contributions are illustrated separately in the bottom; the stress profile in red, the strain anisotropy in blue.*

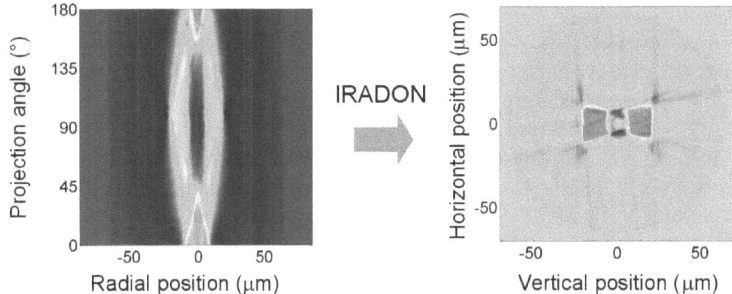

Figure 4.3: *Recovery of the two-dimensional stress profile of a bow-tie fiber (Fibercore HB1250P) from projection data. The inverse Radon transform is used to calculate the stress from the sinogram.*

radial and hoop stress components of the circular symmetric profile. If the total stress profile was directly inserted in equation (2.15) without substracting the virtual "inelastic stress" before, one would get non-physical results, namely non-vanishing stresses outside the fiber cross-section. However, the integral (2.15) yields zero for radial positions outside the fiber, when the stress profile is in accordance with St. Vernant's principle (2.16).

4.2.3 Asymmetric stress profiles: optical tomography

When the stress profile of an optical fiber is no longer cylindrical symmetric, it cannot be recovered from just one single projection. Instead, projection data must be recorded for different fiber orientations by rotating the fiber through a total range of at least 180°. Each projection consists of a fixed number of data points which is determined by the imaging system (CCD) used for data acquisition. The representation of the projection data as a function of the fiber orientation is called "sinogram" and is illustrated for a bow-tie fiber on the left of Fig. 4.3. The projection data are found to vary considerably for the different projection angles. A better understanding of the sinogram is obtained by comparison with the stress profile on the right of Fig. 4.3. For a projection angle of 90°, the fiber is illuminated from "below" (i.e. in parallel to the vertical coordinate), which allows to distinguish the two "bow-tie" regions in the corresponding projection profile. For rotation angles of 0 or 180°, however, one of the "bow-tie" regions is shadowed by the other one, so that no more splitting of the retardation can be observed in the sinogram. Instead, the amount of retardation is almost doubled.

To reconstruct the stress image from projection data as illustrated in

Fig. 4.3, the inverse Radon-transform implementation IRADON of the computer algebra environment MATLAB[1] was used. IRADON uses a filtered backprojection algorithm to perform the inverse Radon transform. In the following, we give a short overview about the theoretical background of the filtered backprojection algorithm. A more detailed description of the filtered backprojection method and its implementations is given in [30].

To account for the orientation of the rotating fiber with respect to the illuminating laser beam, two coordinate systems are introduced. The cartesian coordinate system (x,y) is aligned with respect to the laser beam, with x being parallel and y perpendicular to the ray (see Fig. 4.1). The rectangular coordinate system (u,v), however, is fixed on the fiber. The two coordinate systems are linked by the fiber rotation angle θ according to

$$\begin{bmatrix} u \\ v \end{bmatrix} = \begin{bmatrix} \cos\theta & -\sin\theta \\ \sin\theta & \cos\theta \end{bmatrix} \begin{bmatrix} x \\ y \end{bmatrix} \tag{4.14}$$

For $\theta = 0°$, the two coordinate systems coincide. Equation (4.3) can than be rewritten as a function of rotation angle as

$$\Re(y,\theta) = C \int_{-\infty}^{\infty} \sigma_{zz}(u,v) \, dx \tag{4.15}$$

$$= C \int_{-\infty}^{\infty} \sigma_{zz}(x\cos\theta - y\sin\theta, x\sin\theta + y\cos\theta) \, dx, \tag{4.16}$$

where $\Re(y,\theta)$ is the sinogram of the object to be reconstructed. In a next step, we introduce the two-dimensional Fourier-transformation of the stress profile according to

$$F(\xi,\eta) = C \int_{-\infty}^{\infty}\int_{-\infty}^{\infty} \sigma_{zz}(u,v) \exp[-j(\xi u + \eta v)] \, du dv. \tag{4.17}$$

Here, the rectangular coordinates ξ and η in the frequency domain have been introduced. Conversion of the cartesian into polar coordinates ($\xi = w\cos\theta$, $\eta = w\sin\theta$) yields

$$F(w\cos\theta, w\sin\theta) = C \int_{-\infty}^{\infty}\int_{-\infty}^{\infty} \sigma_{zz}(u,v) \exp[-jw(u\cos\theta + v\sin\theta)] \, du dv \tag{4.18}$$

$$= \int_{-\infty}^{\infty} \Re(y,\theta) \exp(-jwy) \, dy, \tag{4.19}$$

---

[1] http://www.mathworks.com/products/matlab/

where the equality between (4.18) and (4.19) is due to equation (4.16). Thus, $F(\xi,\eta)$ is the Fourier-transform of the sinogram $\Re(y,\theta)$ and can directly be determined as the fast fourier transform (FFT) of the projection data. Finally, the two-dimensional stress profile is found by the inverse Fourier-transformation of equation (4.17) as

$$\sigma_{zz}(u,v) = \frac{1}{4C\pi^2} \int_{-\infty}^{\infty} \int_{-\infty}^{\infty} F(\xi,\eta) \exp[j(\xi u + \eta v))] \, d\xi d\eta \quad (4.20)$$

$$= \frac{1}{8C\pi^2} \int_{0}^{2\pi} \left[\int_{-\infty}^{\infty} F(w,\theta)|w| \exp(jwy) dw \right] d\theta. \quad (4.21)$$

Equation (4.21) is deduced from equation (4.20) by transformation from the rectangular coordinate system (ξ, η) to the polar system (w, θ) in the frequency domain. Further, the property $F(w, \theta + 180°) = F(-w, \theta)$ has been applied. The term in brackets in equation (4.21) represents the filtering operation, where the frequency response of the filter is given by $|w|$. After the filtering operation, the data is backprojected over the image plane for all angles θ. As depicted by equation (4.21), this corresponds to an integration of the filtered backprojections over all angles in the space domain.

The inverse Radon-transform allows to compute two-dimensional axial stress profiles from retardation data. Once the axial stress profile is known, the other components of the stress tensor can be calculated. Therefore, the general procedure is quite similar to the stress profiling in symmetrical fibers using the Abel-inversion. As for the Abel inversion, the pure elastic stress must be separated from the inelastic contribution by applying equation (4.13). Once the elastic stress contibution is identified, the other stress components can be calculated according to the approach described in 2.1.3. As for the Abel-inversion, violation of equation (4.11) leads to non-physical results for the other stress components, particularly to non-vanishing stress outside the fiber.

4.3 Stress measurement setup

4.3.1 Setup description

The stress measurement setup planned and realized within this work is illustrated in Fig. 4.4. The setup basically is assembled from a polarization controller, a Sénarmont compensator, and imaging optics. The polarization controller consists of the combination of a polarizer, a quarter waveplate, and a second rotating polarizer. The first polarizer is illuminated with incoherent light from the laser source. Due to the high spatial coherence of the HeNe-laser, a rotating diffuser was used to impair speckle formation in the image plane. The incoherent light behind the diffuser is collected by a lens (f= 20 mm) and linearly polarized. The linear polarization is converted into circular by a quarterwave plate, whose fast axis is aligned under an angle of 45° with respect to the main axis of the polarizer. The angle of polarization can then be adjusted with the second, rotating polarizer. When the polarization of the incoming light is ideally circular, than the power of the transmitted light does not depend on the rotation angle.

In the following, we use the Jones-matrix formalism to describe the polarization properties of the setup [8]. Alternative methods to describe polarization are the Mueller calculus and the Poincaré-sphere [8]. Both methods are basically superior to the Jones calculus, as they also offer the possibility to describe partially polarized light. Further, the Poincaré sphere is especially convenient for the graphical representation of polarization issues, whereas the Stokes and Mueller calculi are suited to compute the effects of optical polarization on polarized light. We chose the Jones-matrix formalism as we only deal with completely polarized light and due to its simplicity. An alternative description of polarization using the Poincaré sphere was given in [31].

Within the Jones-matrix formalism, the polarization of light leaving the polarization controller is described by

$$\underline{\mathbf{i}} = \begin{bmatrix} \cos(\theta) \\ \sin(\theta) \end{bmatrix}, \tag{4.22}$$

where θ is the rotation angle and the intensity of the beam is normalized to unity. The linearly polarized beam then crosses the a quarterwave plate and the fiber. The fast axis of the quarterwave plate makes an angle of 45° with the fiber. The corresponding Jones-matrix of the waveplate is

$$[\mathbf{QW}] = \frac{1}{\sqrt{2}} \begin{bmatrix} \exp\left(j\frac{\pi}{4}\right) & \exp\left(-j\frac{\pi}{4}\right) \\ \exp\left(-j\frac{\pi}{4}\right) & \exp\left(j\frac{\pi}{4}\right) \end{bmatrix}, \tag{4.23}$$

and the fiber is represented by the Jones-Matrix

$$[\mathbf{FR}] = \begin{bmatrix} \exp(j\delta) & 0 \\ 0 & 1 \end{bmatrix}, \tag{4.24}$$

where δ is the difference in retardation between the fast and the slow axes. The intensity and homogeneity of the light illuminating the fiber is adjusted by a condenser situated between the quarterwave plate and the fiber. The intensity distribution of the illuminated fiber is sampled by a polarization

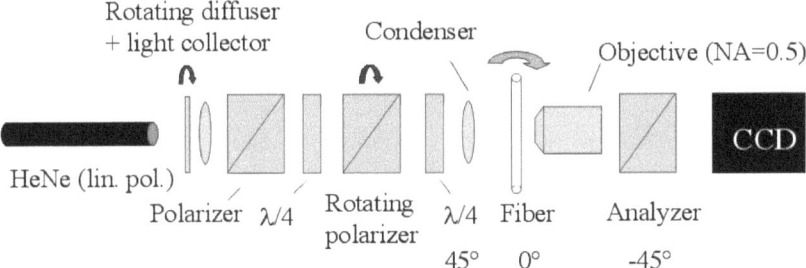

Figure 4.4: *Schematic of the stress measurement setup realized within this work. The polarization optics are illustrated in grey, whereas collimating and imaging optics are violet. A HeNe-Laser serves as light source, and the fiber is imaged on a CCD camera.*

contrast microscope objective, which is especially designed for polarization sensitive measurements and thus does not add any birefringence. The objective images the fiber on a CCD camera, which is used for data acquisition. The combination of quarterwave plate and fiber turns the linear polarization emerging from the rotating polarizer. The light leaving the fiber is again linearly polarized, but its angle is turned in proportion to the retardation δ introduced by the fiber. To convert the change in polarization angle to a modulation of intensity, an analyzer making an angle of $-45°$ with the fiber axis is put between the microscope objective and the CCD camera. The Jones-matrix of the analyzer is

$$[\mathbf{AL}] = \frac{1}{2}\begin{bmatrix} 1 & 1 \\ 1 & 1 \end{bmatrix}. \quad (4.25)$$

The Jones-vector of the light behind the analyzer is found by concatenating the Jones-matrices of all optical elements forming the compensator [8]. One gets the result

$$\underline{o} = [\mathbf{AL}][\mathbf{FR}][\mathbf{QW}]\underline{i} \quad (4.26)$$

$$= \frac{\sqrt{2}}{2}\exp(j\frac{\delta}{2})\left(\cos(\frac{\delta}{2}+\frac{\pi}{4})\cos(\theta) + \cos(\frac{\delta}{2}-\frac{\pi}{4})\sin(\theta)\right)\begin{bmatrix} 1 \\ 1 \end{bmatrix}, \quad (4.27)$$

which represents light polarized under an angle of $-45°$ and with an amplitude depending on both the fiber-induced retardation δ and the angle θ of the light adjusted by the rotating polarizer. The corresponding intensity of

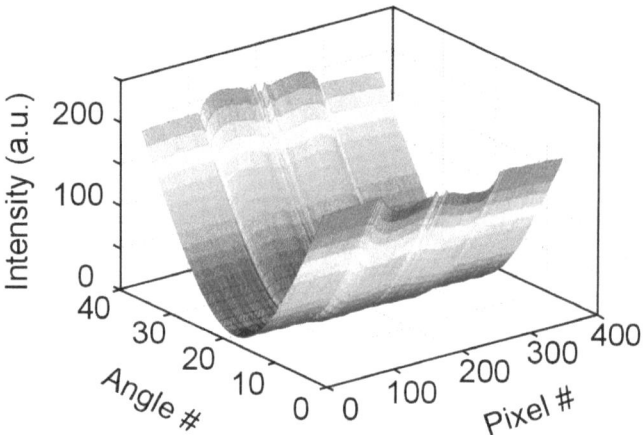

Figure 4.5: *CCD-data captured for the measurement of retardation in an SMF-28® fiber. The intensity is sampled at 36 points over an angle range of ±10°. The retardation is given by the phase of the sine-square dependence of intensity on rotation angle. It is determined with high accuracy by a least square fitting algorithm.*

the light is found by multiplying equation (4.27) with the complex conjugate of its transpose:

$$I = \tilde{\underline{o}}\,\underline{o} = \cos^2(\theta - \frac{\delta}{2} - \frac{\pi}{4}). \tag{4.28}$$

The intensity captured by the CCD camera is thus a sine-square function of the polarizer's rotation angle θ. The fiber-induced retardation δ can directly be deduced from the phase of the sine-square function. The phase can be determined with high precision by fitting a sine-square function to the CCD data. In general, the sine-square function could be fitted to each single pixel of the CCD camera. Thus, the spatial resolution of the system in axial and radial direction would be the same. In practice, however, we average over a certain number of pixels in axial direction to reduce computational time and to improve the signal to noise ratio. The number of vertical pixels taken for averaging is in general adapted to the sample under investigation. If the fiber stress profile is assumed not to change as a function of z, we average over all pixel and get a one-dimensional retardation profile $\Re(y)$. The corresponding pixel-data is illustrated in Fig. 4.5. For each pixel of the y-coordinate, the retardation data is taken at 36 points over a range of ±10°. The intensity varies between 0 and 256 according to the 8-bit encoding of the camera.

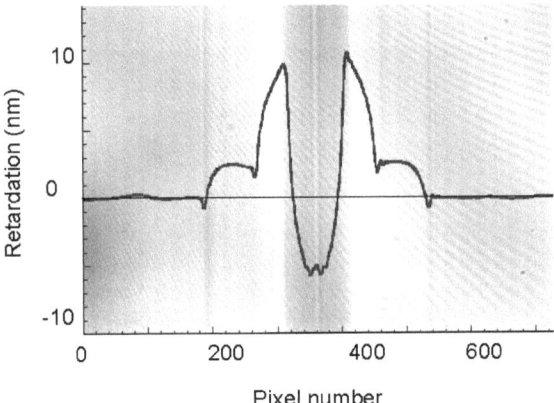

Figure 4.6: *CCD-image recorded for one fixed angle of the polarizer. The intensity modulation is introduced by the fibers retardation profile. The blue line is the highly accurate retardation profile found by fitting a sine-square function to the intensity over all 36 polarizer angles.*

The sine-square behavior of the intensity over the rotation angle θ is obvious from Fig. 4.5. In Fig. 4.6, an image captured from the CCD camera during a measurement procedure is illustrated. The intensity modulation caused by the fiber's retardation profile can clearly be observed. The blue line represents the retardation profile obtained by least square fitting of a sine-square function over all projection angles (cf. Fig. 4.5). For the orientation of the polarizer used to record the CCD-image in Fig. 4.6, the bright regions correspond to a positive retardation, whereas the dark regions are caused by a negative retardation.

Once the retardation profile is determined by least-square fitting of a sine-square function to the CCD data, the corresponding stress profile can be directly computed for cylindrical symmetries using the Abel-inversion (4.10). For asymmetric stress profiles, the fiber has to be turned around its axis, and the next retardation profile has to be taken. Thus, covering a total range of 180°, a sinogran as depicted on the left of Fig. 4.3 is assembled step-by-step, and after the retardation has been determined for the last angle, the stress profile is calculated with the inverse Radon-transform (4.21).

For imaging the fiber on the CCD camera, two different polarization contrast microscope objectives have been used. Their magnifications were 10x and 20x with corresponding numerical apertures of 0.3 and 0.5, respectively. Both objectives are members of the Plan Neofluar family of Carl Zeiss. The 20x objective samples an area of about 260 x 195 μm^2 in the objective plane, the corresponding region for the 10x objective is 520x390 μm^2. The CCD camera (EDH kam02EC) is equipped with a sensor chip of 768 × 574 pixels. The 768 pixels are aligned vertically, in parallel to the y-axis (Fig. 4.1). Consequently, the 574 pixels are aligned along the fiber axis. The resolution of the setup is basically determined by the diffraction-limit of the two objectives used. For the 20x objective, one finds 0.51 μm, for the 10x objective 0.86 μm.

The effects of off-focusing and beam-deflection on the measurement results have been investigated in detail in [32]. It was shown that the combination of the fiber and a 20× microscope objective compensates for a large amount of image distortion. However, for core-cladding index differences in excess of 1%, beam-deflection was found to increase considerably, introducing a significant amount of error in the stress profile. The error can be reduced when the focusing error introduced by a displacement between the fiber center and the objectives focal plane is minimized. Accurate focusing is thus crucial for the precise acquisition of retardation data. Focusing can efficiently be optimized by minimizing the intensity contrast near the fiber core [32].

4.3.2 Setup calibration

To verify the calibration of the setup, stress measurements were performed on a fiber under a well defined strain. The thus obtained stress-strain relationship determines the Young's modulus of the fiber and can be compared with literature values.

The fiber used within the experiment was a hydrogen-loaded SMF-28® standard telecommunication fiber. To determine the strain applied to the fiber, a Bragg grating (λ_B = 1534.79 nm) was written in the fiber using an ArF-laser emitting at 193 nm and the phase mask technique described in subsection 3.4.1. The fiber was clamped between a Teflon plate and a metallic surface with two screws over a total length of 114.55 mm. The fiber was successively strained in steps of 40 μm using a micrometer screw. After each step, the Bragg wavelength as well as the stress profile have been measured. This procedure was repeated 10 times, until the fiber was broken.

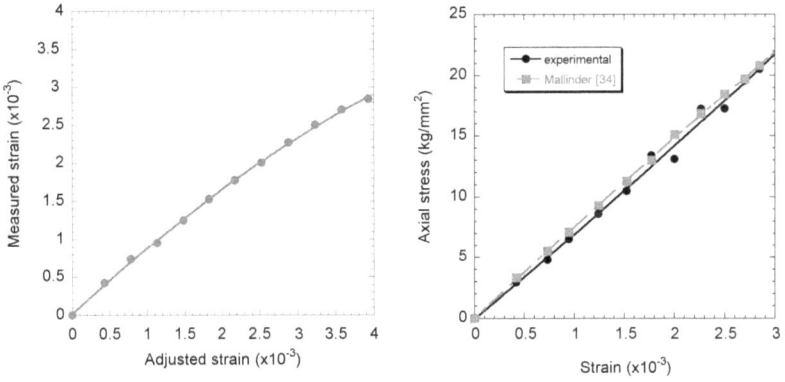

Figure 4.7: Dependence of strain determined from the Bragg-wavelength shift on strain adjusted using the micrometer screw (left). Correlation of measured strain and stress and comparison with literature values (right).

The Bragg wavelength was used to determine the absolute strain value in the fiber. The strain applied to the fiber results in a elongation of the grating period, which causes a wavelength shift according to equation (3.15). The shift is measured with an accuracy of 20 pm using a tunable laser source (Photonetics TUNICS 1550) and an optical powermeter (ANRITSU ML910B) with InGaAs-photodiode (MA9305B). The strain is calculated from the Bragg wavelength shift according to

$$\epsilon_{zz} = \frac{1}{1-p_e} \frac{\Delta \lambda}{\lambda_B} \qquad (4.29)$$

where we introduced the effective photoelastic constant $p_e = 0.21$ [33]. We correlate the strain values set by adjusting the micrometer screw ($\epsilon_{zz} = n \times 40\,\mu\text{m}/114.55\,\text{mm}, n = 0\ldots 10$) with the strain values obtained from the wavelength shift using equation (4.29) in Fig. 4.7 (left). Obviously, the measured strain deviates significantly from the adjusted value. We attribute the difference to a slipping of the fiber due to an insufficient fixation with the two screws. In the following, we use the measured strain values to determine the stress-strain relationship.

For constant stress applied uniformly over the fiber cross section, the integral (4.3) yields

$$\Re(y) = 2C\sqrt{R^2 - y^2}\,\sigma_{zz}. \qquad (4.30)$$

We omit any contribution of inelastic strain to the measured retardation, as it can be considered to be small in comparison to the elastic part for large external forces. The stress can directly be determined from equation (4.30) by inserting the maximum retardation $\Re^{\max}$ in the fiber center ($y = 0$),

$$\sigma_{zz} = \frac{\Re^{\max}}{2CR}. \qquad (4.31)$$

The dependence of stress on measured strain is shown in Fig. 4.7 (right). In general, for large tensile strains, the stress-strain relationship exhibits second order non-linearity according to [34]

$$\sigma_{zz} = E\left(1 + \frac{\gamma}{2}\epsilon_{zz}\right)\epsilon_{zz}, \qquad (4.32)$$

where E is Young's modulus and γ the nonlinearity constant. The experimental values shown in Fig. 4.7 (right) have been fitted to equation (4.31), and the resulting values for E and γ are compiled in Table 4.1 together with the results presented for bulk glasses by Mallinder et al. [34] and for fibers by Bertholds et al. [35]. As the nonlinearity constant can change considerably for only small changes in Young's modulus without strongly degrading the quality of the fit, it has been fixed to the values given in [34, 35]. The corresponding results for Young's modulus deviate by less than 5 % from the results of Mallinder, whereas the difference to the result of Bertholds is almost 10 %. The results presented here have been found only for small elongations, as the fiber broke at the position of the grating for higher stresses. The measurement could basically be improved by using high-strength gratings that exhibit a larger mechanical reliability [36].

E	γ	reference
7.33×10^3 kg/mm^2	5.75	Mallinder et al. [34]
7.08×10^3 kg/mm^2	5.75	this work
6.41×10^3 kg/mm^2	8.66	Bertholds et al. [35]
7.06×10^3 kg/mm^2	8.66	this work

Table 4.1: *Comparison of the stress-strain dependency measured using the setup with literature values*

4.3.3 Accuracy and Repeatability

The accuracy of the measurement is in general defined as the closeness of agreement between the measured and the true value. The difference between the true and the measured value is referred to as the total error of the measurement [37]. The total error is the sum of the systematic (or bias) error and the random (or precision) error. The systematic error is fixed, whereas the random error is given by the repeatability of the measurement. As we have no information about the true stress profile in the fiber, we restrict our error analysis in the following to the assessment of the repeatability. The repeatability is defined as the closeness of agreement between the results of successive measurements of the same measurand carried out under the same conditions of measurement [38]. (The reproducibility, in contrast, is determined from measurements carried out under changed conditions of measurement.)

In [32], error in axial stress introduced by beam deflection was found to vary with the refractive index profile of the specimen. We therefore examine the repeatability in the following for four different germanium-doped fibers (Cabloptic Suisse) differing essentially in their refractive index profile. The properties of the fibers are summarized in table 4.2.

In Fig. 4.8 (above), both the mean value and the standard deviation of the phase retardation (4.2) are illustrated as a function of the radial position for the fiber doped with 9 mol% germanium (preform code 919107.D). Both mean value and standard deviation were calculated for a total number of 10 successive measurements. The standard deviation is found to change considerably as a function of radial position. The higher the gradient of the mean value, the higher is the standard deviation. The maximum standard deviation of 1 nm is found at a radial position of about 15 µm.

The mean value and the standard deviation of the axial stress profile

Preform code	mol% Ge	RI difference	Core diameter
935124.Q	2.9	0.48×10^{-2}	9.4 µm
919107.D	9.0	1.47×10^{-2}	6.7 µm
939068.Q	11.9	1.95×10^{-2}	5.8 µm
939055.Q	18.0	2.95×10^{-2}	3.9 µm

Table 4.2: Germanium-doped fibers (Cabloptic Suisse) used to establish the dependence of the setup's repeatability on the fiber index profile.

calculated from the retardation profile depicted in Fig. 4.8 (above) are shown in the bottom of the figure. In contrast to the retardation error, the error in axial stress is not only caused by the accuracy of the measurement, but also due to the numerical calculation error introduced by the implementation of the Abel-transform (4.10). The increase in standard deviation of the axial stress at the fiber center is basically numerically [39], whereas the local peaks in stress standard deviation at radial positions of about 15, 35 and 60 µm result from the corresponding scattering in the measurement of retardation.

As we are particularly concerned in stress changes of the fiber core, we evaluate in the following the mean standard deviation of both retardation and stress profile within a radius of 5 µm from the core center. The radius was chosen to minimize the influence of error introduced by the retardation gradient at $r \approx 15$ µm for the fiber shown in Fig. 4.8. The results are illustrated in Fig. 4.9. For the mean standard deviation of the retardation, a linear dependence on the index difference of the core is found. The degrading effect of index difference on the accuracy of the measurement reported in [32] can thus be confirmed. For the standard deviation in stress, however, a large scatter around a linear dependence is found. This can be explained by the dependence of the Abel-transform on the derivative of the retardation profile, which makes it extremely sensitive to already small fluctuations in retardation. For a larger number of successive measurements, we thus expect the scattering to decrease.

The standard deviation in the core of an SMF-28® standard telecommunication fiber has been reported to be 0.059 kg/mm² by Park *et al.* in [32]. The refractive index difference of this kind of fiber has about the same value as the Cabloptic fiber doped with 3 mol% germanium, i.e. 5×10^{-3}. However, the standard deviation in our case is about 0.4 kg/mm² and thus more than six times higher than reported by Park. A possible explanation for this differ-

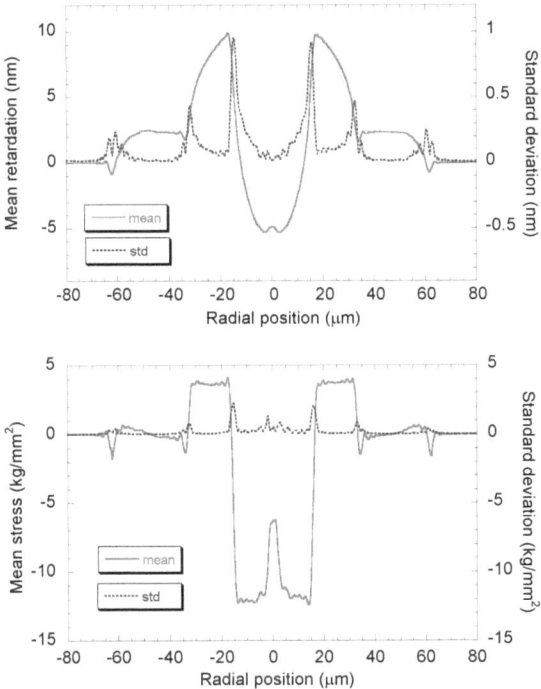

Figure 4.8: Mean value and standard deviation of the phase retardation (above) and the axial stress profile (below) of the fiber doped with 9% germanium.

ence might be the higher stress values and gradients in the Cabloptic-fibers, which lead to a larger error.

4.3.4 Spatial resolution

To gather information about the spatial resolution of the setup, a tomographic measurement on a germanium-doped six-core fiber[2] has been performed. Retardation data were captured as a function of the fiber radial position for 60 rotation angles of the fiber equally spaced by 3°, covering a total range of 180°. The objective used was the Zeiss Plan Neofluar 20×, NA=0.5.

---

[2]The fiber (code number CC_120804_A) was fabricated at the Photonics & Photonic Materials Group, Department of Physics, University of Bath, Bath, England.

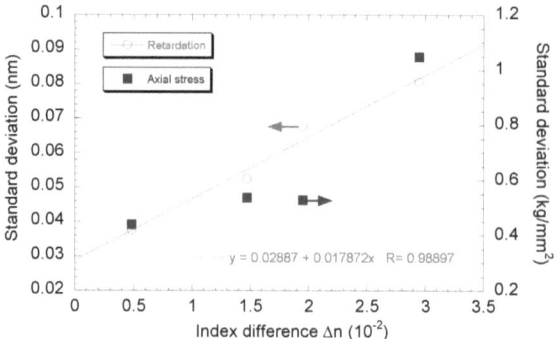

Figure 4.9: *Mean standard deviation of stress and retardation calculated within a radius of 5 µm from the fiber center. The standard deviation of the retardation increases almost linearly with the refractive core index difference, whereas significant scattering from a linear correlation is found for the axial stress.*

In Fig. 4.10, we present a stress image of the whole fiber and a zoom of the tomographic stress profile in the region of the six cores. Five of the six cores were found to have tensile stress maxima of $5.1 \, \text{kg/mm}^2$ while one has a maximum of $6.2 \, \text{kg/mm}^2$, indicating that the fiber has not a perfect six-fold-symmetry. The axial stress in the inner cladding ranges from -2 to $1 \, \text{kg/mm}^2$. The six cores can clearly be resolved spatially and are found to have diameters of about 3.5 ± 0.5 µm. We thus estimate the spatial resolution for the setup to be slightly better than 1 µm.

The star like stress perturbations clearly visible in Fig. 4.10 (above) are artifacts that arise in the "shadow-region" of the tensile cores. They can generally be reduced by increasing the number of projection angles.

The generation of additional synthetic data by numerical interpolation of experimental data can reduce the measurement time considerably. In our case, we obtained almost the same six maximum core stress values for only 20 experimental projection angles by adding 40 interpolated ones. A further increase in the number of experimental projection angles is thus not changing the maximum core stress values considerably.

4.4 Conclusions

A polariscope capable of measuring integrated retardation profiles of optical fiber with high sensitivity and spatial resolution has been developed. The retardation data has been interpreted to be both stress and inelastic strain in-

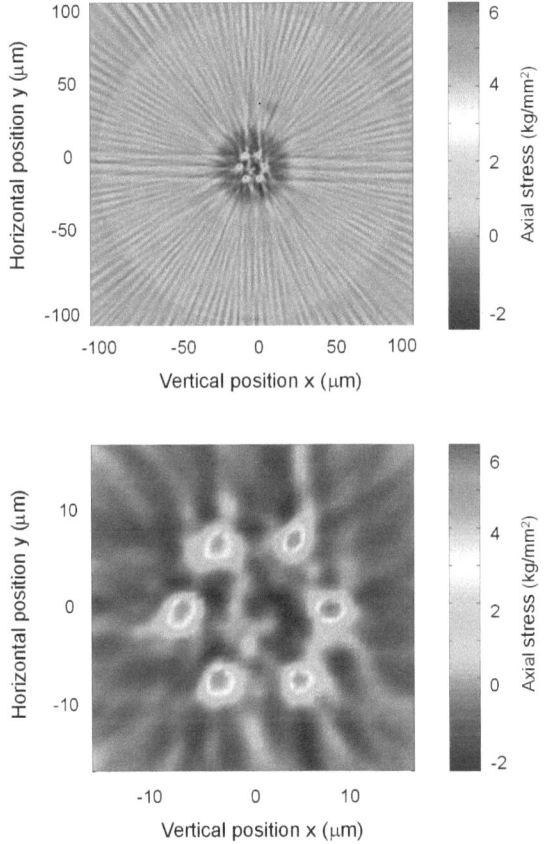

Figure 4.10: Tomographic axial stress profile of the six-core germanium doped-fiber: Overview (above) and zoom of the cores region (below).

duced. The inelastic strain induced retardation can be converted to a virtual "inelastic stress" by applying the Abel-inversion to the overall retardation profile. It can be identified as it is assumed to be constant and as the elastic strain or stress has to be in mechanical equilibrium when no force is applied to the fiber. By rotating the fiber, the setup can also be used to acquire tomographic stress profiles of the specimen. The standard deviation of both retardation and stress has been measured as a function of position and core-cladding refractive index change. They were both found to increase in the core region with rising index change. The spatial resolution was assessed by measuring a six-core fiber and is estimated to be better than 1 μm for the 20× objective.

Bibliography

[1] J. F. Stirling. Frozen strains in glass fibers. *Journal of the Society of Glass Technology*, 39:134–144, 1955.

[2] P. M. Sutton. Stress measurements in circular cylinders. *Journal of the American Ceramic Society*, 41(3):103–109, 1958.

[3] D. A. Krohn. Determination of axial stress in clad glass fibers. *Journal of the American Ceramic Society*, 53(9):504–507, 1970.

[4] U. C. Paek and C. R. Kurkjian. Calculation of cooling rate and induced stresses in drawing of optical fibers. *Journal of the American Ceramic Society*, 58(7–8):330–335, 1975.

[5] M. J. Saunders. Determination of the stress in optical fibers by means of a polariscope. *Review of Scientific Instruments*, 47(4):496–500, 1976.

[6] K. Nagano, S. Kawakami, and S. Nishida. Change of the refractive index in an optical fiber due to external forces. *Applied Optics*, 17(13):2080–2085, 1978.

[7] N. Shibata, K. Jinguji, M. Kawachi, and T. Edahiro. Nondestructive structure measurement of optical-fiber preforms with photoelastic effect. *Japanese Journal of Applied Physics*, 18(7):1267–1273, 1979.

[8] P. S. Theocaris and E. E. Gdoutos. *Matrix Theory of Photoelasticity*. Springer Series in Optical Sciences, 1979.

[9] N. Lagakos, R. Mohr, and O. H. El Bayoumi. Stress optic coefficient and stress profile in optical fibers. *Applied Optics*, 20(13):2039–2313, 1981.

[10] P. L. Chu and T. Whitbread. Measurement of stresses in optical fibre or preform. *Electronics Letters*, 18(1):28–29, 1982.

[11] P. L. Chu and T. Whitbread. Measurement of stresses in optical fiber and preform. *Applied Optics*, 21(23):4241–4245, 1982.

[12] R. S. Anderssen and R. B. Calligaro. Non-destructive testing of optical fiber preforms. *Journal of the Australian Mathematical Society B*, 23(2):127–135, 1981.

[13] R. B. Calligaro, D. N. Payne, R. S. Anderssen, and B. A. Ellem. Determination of stress profiles in optical-fibre preforms. *Electronics Letters*, 18(11):474–475, 1982.

[14] P. K. Bachmann, W. Hermann, H. Wehr, and D. U. Wiechert. Stress in optical waveguides. I. Preforms. *Applied Optics*, 25(7):1093–1098, 1986.

[15] P. K. Bachmann, W. Hermann, H. Wehr, and D. U. Wiechert. Stress in optical waveguides. II. Fibers. *Applied Optics*, 26(7):1175–1182, 1987.

[16] P. Ledoux, J. Auge, L. Y. Boniort, P. Dupont, J. Goudeau, J. M. Saugrain, J. C. Rousseau, and Y. Mohanna. Stress measurements of fibers and preforms. *Journal of Lightwave Technology*, 7(8):1270–1274, 1989.

[17] Y. Mohanna, J. M. Saugrain, J. C. Rousseau, and P. Ledoux. Relaxation of internal stresses in optical fibers. *Journal of Lightwave Technology*, 8(12):1799–1801, 1990.

[18] T. Rose, D. Spriegel, and J. R. Kropp. Fast photoelastic stress determination: application to monomode fibres and splices. *Measurement Science & Technology*, 4(13):431–434, 1993.

[19] T. Abe, Y. Mitsunaga, and H. Koga. Novel measurement method for axial residual stress in optical fibre. *Electronics Letters*, 21(1):4–5, 1985.

[20] T. Abe, Y. Mitsunaga, and H. Koga. Photoelastic computer tomography: a novel measurement method for axial residual stress profile in optical fibres. *Journal of the Optical Society of America A*, 3(1):133–138, 1986.

[21] A. E. Puro and K. J. Kell. Complete determination of stress in fiber preforms of arbitrary cross section. *Journal of Lightwave Technology*, 10(8):1010–1014, 1992.

[22] Y. Park, U. C. Paek, and D. Y. Kim. Complete determination of the stress tensor of a polarization-maintaining fiber by photoelastic tomography. *Optics Letters*, 27(14):1217–1219, 2002.

[23] Y. Park, U. C. Paek, and D. Y. Kim. Determination of stress-induced intrinsic birefringence in a single-mode fiber by measurement of the two-dimensional stress profile. *Optics Letters*, 27(15):1291–1293, 2002.

[24] H. Poritsky. Analysis of thermal stresses in sealed cylinders and the effect of viscous flow during annealing. *Physics*, 5:406–411, 1934.

[25] R. C. O'Rourke. Three-dimensional photoelasticity. *Journal of Applied Physics*, 22(7):872–878, 1941.

[26] C. J. Dasch. One-dimensional tomography: a comparison of Abel, onion-peeling, and filtered backprojection methods. *Applied Optics*, 31(8):1146–1152, 1992.

[27] N. H. Abel. Auflösung einer mechanischen Aufgabe. *Journal für die reine und angewandte Mathematik*, 1:153–157, 1826.

[28] Dietrich Marcuse. *Theory of Dielectric Optical Waveguides*. Academic Press, INC., second edition, 1991.

[29] Y. Park, U. C. Paek, S. Han, B. H. Kim, C. S. Kim, and D. Y. Kim. Inelastic frozen-in stress in optical fibers. *Optics Communications*, 242(4–6):431–436, 2004.

[30] A. C. Kak and M. Slaney. *Principles of Computerized Tomographic Imaging*. Society of Industrial and Applied Mathematics SIAM, 2001.

[31] Y. Park, T. J. Ahn, Y. H. Kim, W. T. Han, U. C. Paek, and D. Y. Kim. Measurement method for profiling the residual stress and the strain-optic coefficient of an optical fiber. *Applied Optics*, 41(1):21–26, 2002.

[32] Y. Park, S. Choi, U. C. Paek, K. Oh, and D. Y. Kim. Measurement method for profiling the residual stress of an optical fiber: detailed analysis of off-focusing and beam-deflection effects. *Applied Optics*, 42(7):1182–1190, 2003.

[33] A. Iocco. *Tunable Fiber Bragg Grating Filters*. PhD thesis, Ecole Polytechnique Fédérale de Lausanne, 1999.

[34] F. P. Mallinder and B. A. Proctor. Elastic constants of fused silica as a function of large tensile strain. *Physics and Chemisty of Glasses*, 5(4):91–103, 1964.

[35] A. Bertholds and R. Dändliker. Deformation of single-mode optical fibers under static longitudinal stress. *Journal of Lightwave Technology*, 5(7):895–900, 1987.

[36] D. Varelas, D. M. Costantini, H. G. Limberger, and R. P. Salathé. Fabrication of high mechanical resistance Bragg gratings in single-mode optical fibers with continuous-wave ultraviolet laser side exposure. *Optics Letters*, 23(5):397–399, 1998.

[37] R. Barlow. *Statistics : a guide to the use of statistical methods in the physical sciences*. The Manchester physics series, 1989.

[38] Barry N. Taylor and Chris E. Kuyatt. NIST Technical Note 1297 - Guidelines for Evaluating and Expressing the Uncertainty of NIST Measurement Results. Technical report, National Institute of Standards and Technology, United States Department of Commerce, 1994.

[39] D. Marcuse and H. M. Presby. Focusing method for nondestructive measurement of optical fiber index profiles. *Applied Optics*, 18(1):14–22, 1979.

Chapter 5

Laser-induced stress changes in optical fibers

In this chapter, we present results which were obtained using the measurement setup described in the previous chapter. In the first section, we describe stress changes observed in germanium-doped fibers due to femtosecond-laser[1] and 244 nm continuous-wave irradiation[2]. For both fibers, asymmetrical stress profiles were found, so that tomographic measurements had to be performed. The fs-laser irradiated fiber is a standard SMF-28® telecommunication fiber, whereas the UV-irradiated fiber is a highly germanium-doped fiber manufactured by FORC[3]. For the highly germanium-doped fiber, stress and strain changes due to thermal annealing have also been investigated.

In the second section of this chapter, we summarize results obtained for phosphorus- and nitrogen-doped fibers[4], that where drawn at different drawing tensions. We assess the dependence of stress and inelastic strain on the fiber drawing tension. In a next step, stress changes due to irradiation with pulsed Excimer-irradiation at 193 nm are presented and related to the irradiation conditions. The phosphorus-doped fibers were hydrogen-loaded before irradiation, whereas the nitrogen-doped fibers were pristine.

In the third and terminating section, we summarize and discuss the results of the two preceding sections. The contribution of compaction and the accompanying photoelastic response are estimated for the different dopants and irradiation conditions.

---

[1] results published in [1]
[2] results published in [2]
[3] Fiber Optics Research Center at the General Physics Institute of the Russian Academy of Science, Moscow, Russia.
[4] results published in [3]

5.1 Germanium-doped fibers

5.1.1 Femtosecond-laser induced stress changes

Introduction

Interaction of ultra-short laser pulses with transparent optical solid matter has been studied extensively in the last thirty years [4, 5]. The refractive index modification of the irradiated glass depends strongly on material, focusing, as well as laser parameters. Laser and focusing parameters determine the pulse intensity and peak power, which can be related to the threshold intensity for optical breakdown and the critical power for self-focusing in the material. If the intensity of the pulse exceeds the threshold intensity, strong plasma formation is observed, leading to a permanent damage of the glass volume. Exceeding the critical power results in self-focusing and laser-pulse filamentation [6].

Recently, femtosecond lasers have been used to write three-dimensional waveguide structures in a variety of bulk glass samples [7, 8]. Furthermore, long period gratings [9] as well as fiber Bragg gratings [10] have been realized by modification of the core index in germanosilicate fibers. For waveguide as well as grating formation, intensity and power of the femtosecond pulses are in general below their corresponding threshold values for optical breakdown and self-focusing. Thus, no losses are introduced and the waveguide shape can more accurately be controlled.

The underlying mechanism of the refractive index change in the sub-breakdown intensity region has not yet been completely identified. Raman data indicate an increase in the number of 4- and 3-membered ring structures for fused silica, which is associated with a densification of the glass after exposure to the femtosecond irradiation [11]. For germanosilicate fibers, it has already been pointed out earlier by Dianov *et al.* that UV-irradiation changes the Raman spectra of the fiber [12], indicating a densification of the doped core region. In the case of germanosilicate fibers, the densification is accompanied by a modification of the fiber residual stress profile [13]. An altered stress profile should thus also be found in optical fibers after interaction with femtosecond laser pulses.

Experiment

Within the scope of this work, long period gratings have been employed for the stress measurements, as they provide irradiated as well as non-irradiated sections, and their spectrum further allows to estimate the induced refractive

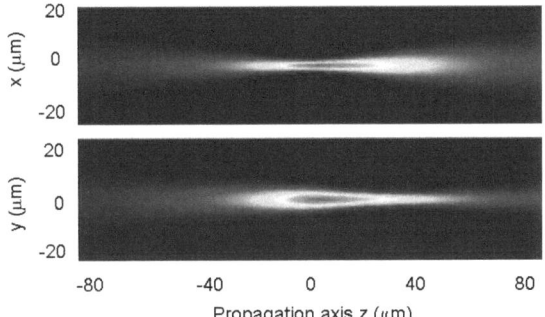

Figure 5.1: *Intensity distribution for the focusing-conditions used within this experiment. The intensity is given by equation (5.1), the astigmatism introduced by the circular shape of the fiber can clearly be observed. The upper figure gives the intensity in the plane $y = 0\,\mu m$, the lower figure for $x = 0\,\mu m$.*

index change. The long period gratings have been written at USTL[5] in standard SMF-28® telecommunication fiber. Laser pulses of 160 fs duration and 0.27 µJ energy were generated from a Ti^{3+}:Al_2O_3 regenerative amplifier (Coherent RegA9000) at a wavelength of 800 nm and a 200 kHz repetition rate. The fiber was longitudinally translated with respect to the femtosecond laser beam at a speed of 2.67 mm/s. The laser beam was interrupted periodically with an electronic shutter, resulting in a grating of 40.05 mm length with 450 µm period and duty cycle of 0.5. A polarizer ensured a beam polarization perpendicular to the fiber axis. Maximization of the guided fluorescence light in the fiber using a photomultiplier optimized the alignment of the fiber core with respect to the focus of the radiation.

The pulses were weakly focused onto the fiber core with a 5× microscope objective ($NA = 0.1$). The pulse peak power is estimated to be $P_{\max} = 0.94\,E_p/\tau_p = 1.59\,\text{MW}$, where τ_p is the pulse-width (FWHM) of a Gaussian-shaped pulse of energy E_p. The peak power is about four times smaller than the critical power for self-focusing in silica [5], but no values have been reported to our knowledge for the critical power in germanosilicate glass. Above the critical power, catastrophic collapse of the laser beam to a singularity is predicted [4]. Supercontinuum generation and collapse due

---

[5]Laboratoire de Physique des Lasers, Atomes et Molécules, Université des Sciences et Technologies de Lille, Lille, France.

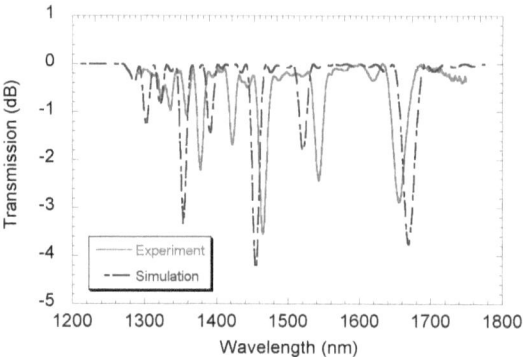

Figure 5.2: *Comparison of the LPG spectrum with simulation data. The laser-induced index change is determined to be about 4×10^{-4}.*

to self-focusing have been reported to occur above the same power threshold [14]. During the writing process, no supercontinuum generation could be observed by eye in the irradiated regions. We thus conclude that self-focusing is still weak in our case and model the incoming laser intensity to be Gaussian

$$I(x,y,z) = \frac{P_{\max}}{4w_y(x)w_z(x-x_0)} \times \exp\left[-2\left(\frac{y^2}{w_y^2(x)} + \frac{z^2}{w_z^2(x-x_0)}\right)\right], \quad (5.1)$$

where the transverse beam dimensions along the laser beam are given by $w_i(x') = w_{0i}[1 + (x'/x_{Ri})^2]^{0.5}$, $i = y, z$, with beam waists w_{0i} and Rayleigh ranges $x_{Ri} = \pi(w_{0i}^2/\lambda)$ perpendicular ($i = y$) and parallel ($i = z$) to the fiber axis. The offset distance x_0 between the two beam waists reflects the astigmatism introduced by the curvature of the fiber in y-direction. For our focusing conditions, we find $w_{0y} = 2.5\,\mu m$, $w_{0z} = 1.8\,\mu m$, and $x_0 = 28.1\,\mu m$. Due to the astigmatism, the two foci do not coincide. The maximum intensity of the beam occurs between them and is found for a peak power of $P_{\max} = 1.59\,MW$ and with equation (5.1) to be $I_{\max} = 6 \times 10^{12}\,W/cm^2$. The intensity distribution according to equation (5.1) is illustrated in Fig. 5.1 for the two planes $y = 0\,\mu m$ (above) and $x = 0\,\mu m$ (below).

To measure the two-dimensional stress profile, the axial stress induced phase retardation profile of the fiber is determined for 60 projection angles from $0°$ to $180°$ using the setup described in the previous chapter. The axial stress profile is then calculated by using the inverse Radon transformation equation (4.21) of the projection data. The axial stress distribution σ_{zz} is then used to determine the other components (σ_{rr}, $\sigma_{\theta\theta}$, $\sigma_{r\theta}$) of the stress tensor completely as described in section 2.1.3. Photoelasticity relates the stress tensor to the refractive index tensor according to equation (3.5), so the stress-induced component of the total refractive index change can be evaluated using the stress measurements.

Results

The spectrum of a long period grating obtained using the aforementioned writing conditions is illustrated in Fig. 5.2. The maximum rejection strength is about 3.5 dB, and the excess loss is about 0.3 dB. Due to the good quality of the measured grating spectrum [15], comparison with simulated spectra could be used to estimate the induced index change. Experimental results and synthetic data from grating modeling match well for an induced refractive index change of $(4 \pm 0.2) \times 10^{-4}$. The modeling is based on a refractive index that is increased over an azimuthal angle of $0 < \phi < 120°$ within the irradiated region of the fiber core to account for a misalignment of the illuminating beam with respect to the fiber center. The modeled spectra is also included in Fig. 5.2. We attribute the observed differences between experimental and synthetic data essentially to our simplifying assumption for the asymmetry in core index change.

Figure 5.3 shows the two-dimensional total axial stress profile (including the virtual "inelastic stress") of the fiber after irradiation with the femtosecond laser pulse train. The x-axis is the axis parallel to the incident laser beam. The stress is increased over an almost elliptical region in the core to a peak value of 5.2 kg/mm^2. The core region has a diameter of 8.2 µm and can be identified in Fig. 5.3 and 5.4 by its negative axial stress. In Fig. 5.4, the two cross-sections parallel (a) and perpendicular (b) to the incident beam are illustrated and compared to the stress profile in the non-irradiated region. The maximum stress change in the core is found to be about 6.2 kg/mm^2. A small misalignment of the stress-modified region with respect to the core center can be observed in Fig. 5.4(b). The stress is increased over a length of 8.5 µm (FWHM) and a width of 2.5 µm.

The virtual "inelastic stress" is determined according to equation (4.13) and yields 0.56 kg/mm^2, corresponding to a strain anisotropy of 6.16×10^{-5}. The strain anisotropy is thus slightly higher as for the SMF-28® fiber presented in the Fig. 4.2 of the last chapter. The SMF-28® fiber presented in Fig. 4.2, however, is not identical to the one presented here. We thus conclude that the residual stress and strain profiles for different SMF-28® fibers can be different.

To obtain the two transverse components of the photoelastic index change [16], the stress tensor has been determined completely according to the procedure described in section 2.1.3. This allows separating the photoelastic

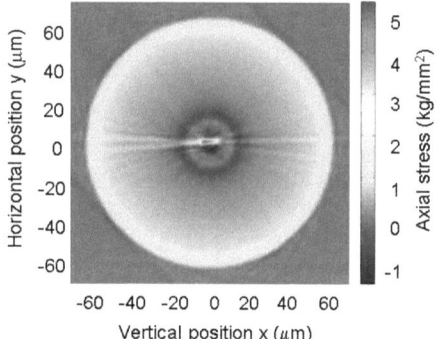

Figure 5.3: Two-dimensional axial stress profile of SMF-28® fiber after irradiation with femtosecond-pulses. The fiber was irradiated along the x-axis. The peak laser-induced stress is $5.2\,\text{kg/mm}^2$ in the core region.

contribution from the total refractive index change [17]. Figure 5.5(a) shows the stress-induced index change for the x-component, Fig. 5.5(b) for the y-component of the refractive index change tensor. For both components, the index change is negative due to the overall tension increase, and peaks at -3.7×10^{-4} and -4.7×10^{-4}, respectively. The modulus of photo-elastic index change is thus, as for UV-illuminated germanoslicate fibers [17], of the same order of magnitude as the total index increase. The inelastic contribution of the total index change must have a magnitude twice as large as the net index change.

For both components, the photoelastic index change in the middle of the modified region is less pronounced than in the adjacent regions. A possible explanation is the lower Ge-concentration in the center of the core region due to the preform fabrication process.

Discussion

To compare the laser intensity profile with the stress changes shown in Fig. 5.4, FWHM-values of the intensity can be calculated from equation (5.1) for both x- and y-direction. In x-direction, we get a length of 47 μm, in y-direction a width of 4.2 μm. As mentioned above, the FWHM-values for the stress change have been found to be 8.5 μm and 2.5 μm, respectively. If the change in residual stress only depends on intensity, the modified region should thus be more than twice as long. The length of the stress-modified

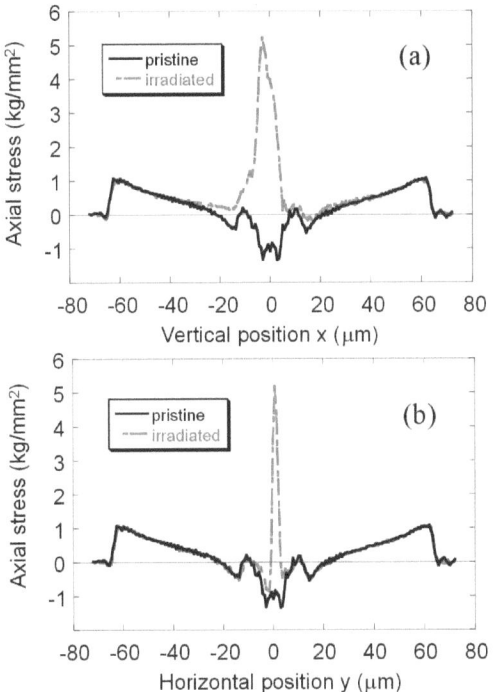

Figure 5.4: *Axial stress profiles parallel (a) and perpendicular (b) to the incident femtosecond laser pulse train. In the focus region, the peak stress increase is $6.2\,\text{kg/mm}^2$.*

region is rather limited to the extension of the fiber core. We thus conclude that the modification in core stress is strongly supported by the germanium-doping of the core region.

For germanosilicate fibers, a linear dependence of refractive index change on axial stress change, $\Delta n = (0.8 \pm 0.2) \times 10^{-4} \Delta \sigma_{zz}$, has been reported after irradiation with pulsed UV-light [13]. For a measured peak axial core stress increase of $6.2\,\text{kg/mm}^2$ in the case of femtosecond-laser irradiation, this corresponds to an index increase of $(4.96 \pm 1.24) \times 10^{-4}$. As mentioned above, a refractive index change of 4×10^{-4} had been found by modeling the transmission spectrum. The two values thus agree well within error, which might indicate that the nature of the glass change introduced by UV- and femtosecond laser irradiation is similar in our case, where low energy pulses were weakly focused into the fiber.

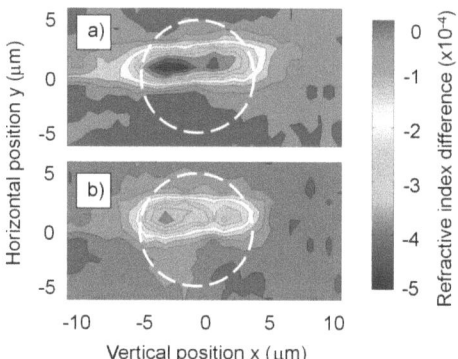

Figure 5.5: Stress-induced change for the x- (a) and y- (b) component of the refractive index change tensor. The circle indicates the Ge-doped core region.

The similarities between the changes in glass properties reported here and for UV-irradiation suggest that the underlying mechanisms may be similar. The UV-induced index change in germanosilicate fibers is triggered by photo-bleaching of Ge oxygen-deficient centers through two-photon absorption [18]. The emerging electron is re-trapped at the original site or at some other defect site. As a result, the shape of the molecule is reconfigured, changing the absorption as well as the density of the material. For irradiation with 240 nm pulses, index changes up to 1.2×10^{-3} have been reported in standard telecommunication fiber [19].

We suggest that for the femtosecond induced index changes reported here, photo-ionization is also the triggering mechanism. For 800 nm light, five photons are necessary to bridge the 7.2 eV band gap of germanium-doped silica [18]. Once the corresponding electrons have been shifted to the conduction band, the recombination mechanisms might be similar to the case of UV photo-ionized glasses.

Conclusion

In summary, we have directly measured core stress changes in Ge-doped fibers irradiated with weakly focused femtosecond laser pulses. The peak change in refractive core index was estimated to be 4×10^{-4} by comparing the LPG spectra with simulation data. The stress was found to increase by up to 6.2 kg/mm² in the focus. Similarities to stress and index changes observed in UV-irradiated fibers suggest that in both cases, the underlying mechanisms in the Ge-doped region are similar. The main difference is the number of photons needed to trigger the mechanisms by electron excitation.

5.1.2 Annealing-induced stress changes in UV-irradiated germanium-doped fibers

Introduction

Thermal annealing experiments of UV-written Bragg gratings serve as a means to gather information about the composition of the index change from distributions with different thermal stability and thus allow predictions about the long-term stability of refractive index changes [20, 21]. In addition, annealing experiments can be used for accelerated aging tests of optical devices. Stress measurements serve as a means to identify the compaction-induced index change triggered by UV-irradiation [17]. Here, we combine the two techniques to evaluate the thermal stability of the UV-induced compaction and its contribution to the overall refractive index change in fiber gratings.

Demarcation mapping theory

The stress measurement presented in this section have been correlated to changes in refractive index that were recorded during a continuous annealing of the samples [22]. The annealing behaviour has been described in terms of demarcation energy [20]. The demarcation energy depends both on time t and temperature T and is defined as

$$E_d = k_B T \ \ln(\nu_0 t), \tag{5.2}$$

where k_B is the Boltzmann-constant and ν_0 the attempt frequency, which has to be determined experimentally. The characterization of the annealing solely in terms of demarcation energy and not in terms of both time and temperature allows to predict long-time annealing behaviour by observing annealing at shorter times and elevated temperature;

$$E_d = k_B T_1 \ \ln(\nu_0 t_1) = k_B T_2 \ \ln(\nu_0 t_2). \tag{5.3}$$

The demarcation energy is interpreted as the energy below which the localized states have been depopulated as illustrated in Fig. 5.6. With an energy density function $\tilde{g}(E)$ of occupied states at $t = 0$ and a normalized mean occupation number

$$\tilde{f} = \exp\left[-\nu_0 t \ \exp\left(-\frac{E}{k_B T}\right)\right], \tag{5.4}$$

the total number of occupied states is given by the overlap integral [20]

$$N(t) = \int\limits_0^\infty \tilde{g}(E)\tilde{f}dE. \tag{5.5}$$

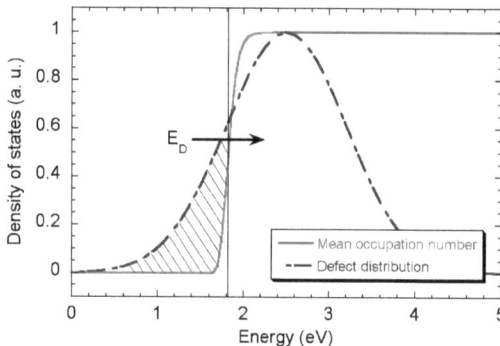

Figure 5.6: *Thermal depopulation of the density of occupied states. All states with an energy lower than the demarcation energy E_d have been thermally released.*

Due to the definition of the demarcation energy, the mean occupation number (5.4) can be approximated by

$$\tilde{f} = \exp\left[-\nu_0 t \exp\left(-\frac{E}{k_B T}\right)\right] \cong \begin{cases} 0, & E < E_D, \\ 1, & E > E_D. \end{cases} \quad (5.6)$$

The total number of occupied states can thus be given as a function of demarcation energy according to

$$N(E_d) \cong \int_{E_d}^{\infty} \tilde{g}(E)\, dE. \quad (5.7)$$

In Fig. 5.6, the hatched region indicates the number of occupied states that have already been thermally released, as their energy is below the demarcation energy. The demarcation energy advances to the right with both time and rising temperature and thus permanently reduces the total number of populated states. Once the total number of occupied states has been determined as a function of demarcation energy, the initial density of occupied states $\tilde{g}$ can be found by differentiation.

In grating annealing experiments, the total number of occupied states is considered to be proportional to the gratings modulation index amplitude. The index decay is then recorded for different annealing conditions and plotted as a function of demarcation energy according to equation (5.2). The correct decay attempt frequency ν_0 causes all annealing data to fall on a single line, the so-called "master curve". Obviously, on needs at least two data sets (i.e. two different temperatures for isothermal annealing, two different times intervals for isochronal annealing, or two temperature ramps for annealing with a temperature sweep) to obtain the "master curve". Once the "master curve" is identified, the distribution of occupied states can be determined by differentiating equation (5.7).

The concept of demarcation mapping does not only apply to UV-induced fiber gratings, but to thermally activated processes in general [23]. Here, we want to adopt it to the case of stress relaxation. Therefore, stress changes would have to be determined for at least two temperature sweeps, and the attempt frequency would have to be identified in the same way as for refractive index changes. However, as the amount of measurement time already for one temperature sweep is considerable, we only acquired data for one single temperature ramp. The data was then correlated with the index data as a function of demarcation energy. This demarcation energy, however, does strictly only apply to the refractive index and not to the stress annealing.

Experiment

The fiber under investigation has been fabricated by MCVD technique at FORC and is doped with 14 mol% germanium in the core. A Lloyd interferometer was used to write Bragg gratings at five different total doses with 244 nm cw-irradiation and a writing intensity of $50\,\mathrm{W/cm^2}$. The annealing of the UV-induced index change was characterized using the technique described in [21, 22]. The demarcation energy for a temperature sweep can be calculated by rearranging equation (5.3) and dividing the sweep into a discrete number of time intervals with corresponding temperature. The attempt frequency was determined by comparison of annealing curves measured for three different temperature sweeps and found to be $\nu_0 = 10^{13}\,\mathrm{Hz}$. The at-

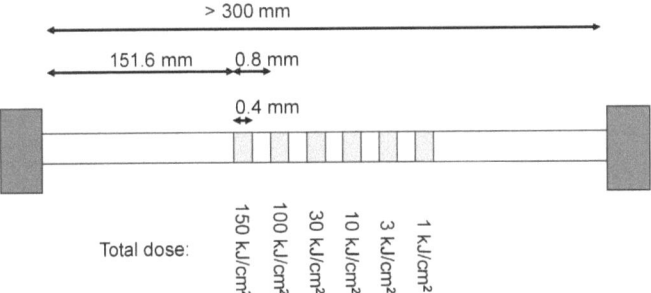

Figure 5.7: The fiber samples for the annealing experiment have been irradiated homogeneously with six different total doses.

tempt frequency did not change with the total irradiation dose, whereas the shape of the master curve was found to depend on the exposure strongly.

In addition, six fiber samples have been prepared for stress measurements at FORC. As illustrated in Fig. 5.7, six regions of 0.4 mm in length, separated by five regions of equal length have been irradiated homogeneously with total doses of 1, 3, 10, 30, 100, and 150 kJ/cm² in each sample. Five of the six fiber samples have subsequently been annealed in a computer-controlled furnace, using a linear temperature sweep of 0.25 K/s. The annealing was stopped when the samples had reached temperatures of 496, 827, 992.5, 1158, and 1323 K, corresponding to refractive index demarcation energies of 1.5, 2.5, 3, 3.5, and 4 eV, respectively.

The two-dimensional stress profiles were measured for seven different projection angles, equally spaced by 30° and covering a total range of 0 - 180°. To improve the quality of the reconstructed images, projection data between the measured data were interpolated.

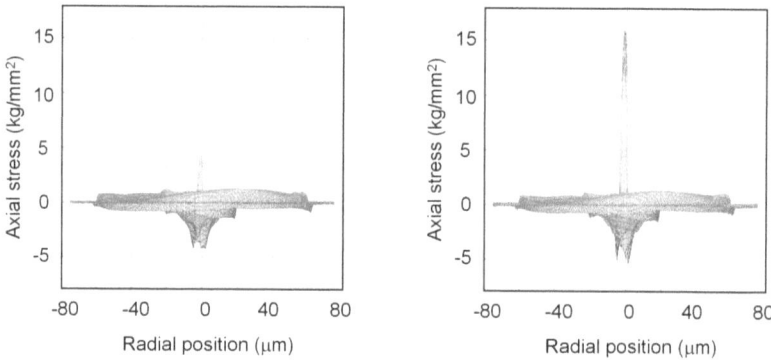

Figure 5.8: Axial two-dimensional stress profile of the pristine fiber (left) and core stress increase after irradiation with a total dose of 150 kJ/cm² (right).

Results and Discussion

In Fig. 5.8, the tomographic axial stress profile of the pristine fiber is presented on the left, and can be compared to the stress profile after irradiation with 150 kJ/cm² on the right. For both stress profiles, stress asymmetries are found particularly in the cladding. The axial core stress increases considerably from 4.4 kg/mm² up to 16 kg/mm² after UV-irradiation. For all samples under investigation, tomographic stress profiles as shown in Fig. 5.8

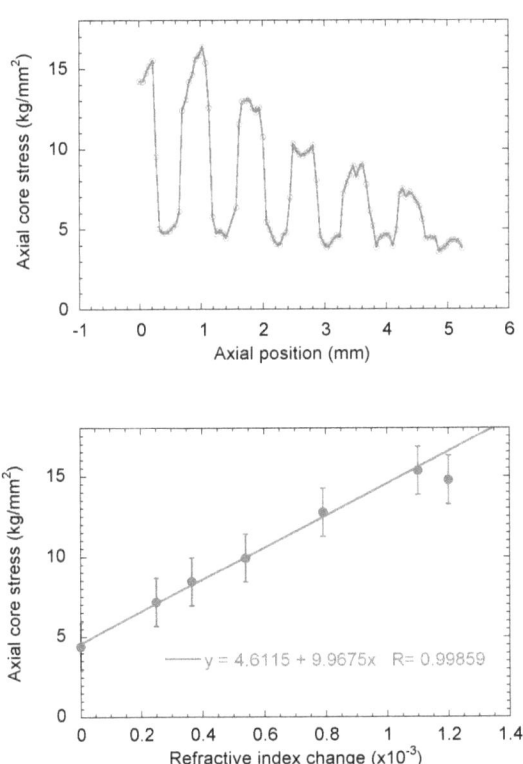

Figure 5.9: Evolution of core stress as a function of axial position for the non-annealed sample (above) and the corresponding linear correlation of stress with index change (below).

were captured along the fiber with an axial resolution of 56 µm. The resulting axial core stress distribution is shown in the upper part of Fig. 5.9 for the non-annealed sample. The higher the total dose, the higher is the increase in axial core stress in the respective region. For the region irradiated with the highest total dose ($z = 0$ in Fig. 5.9), the core stress increase appears to saturate. In the lower part of Fig. 5.9, the increase in axial core stress is correlated with the change in refractive index. As reported in [13] for pulsed irradiation, a linear relation between stress and index change is found, expect for the highest irradiation dose. The slope of the linear fit to the data (excluding the highest irradiation dose) is $\sim 0.9 \times 10^4$ kg/mm^2, which is within the range of $10^4/(0.8 \pm 0.2)$ kg/mm^2 found in [13].

The changes in axial core stress can be represented as a function of modulation index demarcation energy. Therefore, we distinguish between UV-induced core stresses and residual core stress introduced into the fiber during fabrication [24]:

$$\sigma_{Co}^{tot}(E_d) = \sigma_{Co}^{res}(E_d) + \sigma_{Co}^{UV}(E_d), \tag{5.8}$$

where E_d is the index demarcation energy of the respective sample. The evolution of residual core stress with index demarcation energy was obtained by observing the non-irradiated parts of the fiber and is shown on the upper part of Fig. 5.10. The increasing stress for demarcation energies $E_d > 3$ eV is explained by the annealing of drawing induced compressive core stresses. The annealing of the virtual "inelastic stress" component representing the drawing-induced strain anisotropy is depicted on the lower part of Fig. 5.10. Up to demarcation energies of about 3.3 eV the strain anisotropy is constant and falls down for higher energies. The strain anisotropy thus shows higher resistance to annealing in comparison to the drawing induced stress illustrated on the upper part of Fig. 5.10.

Internal stress relaxation in optical fibers have been investigated by Mohanna et al. [25] for isothermal annealing experiments. The annealing can be described by the empirical equation

$$\sigma/\sigma_0 = \exp\left[-\exp\left(-q - C/T\right)t\right]^b, \tag{5.9}$$

where σ_0 is the initial stress value, $q = -19.7$, and $C = 2.45 \times 10^4$ K. The exponential factor b is assumed to have a constant value of 0.63 [25]. For the case of temperature sweeps, we have modified equation (5.9) by dividing the time into discrete intervals of length Δt and assuming a temperature step of ΔT between two subsequent time intervals. Equation (5.9) than reads

$$\sigma(N\,\Delta t)/\sigma_0 = \prod_{i=1}^{N} \exp\left[-\exp\left(-q - C/T_i\right)\Delta t\right]^b, \tag{5.10}$$

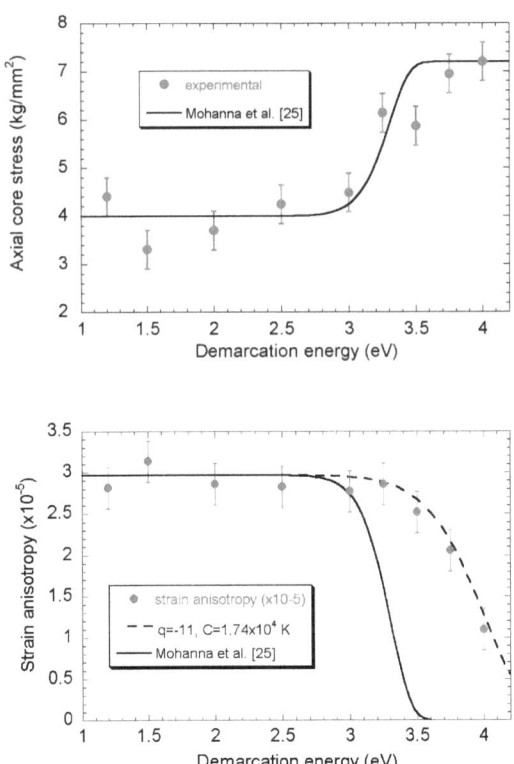

Figure 5.10: Residual axial core stress as a function of demarcation energy for the pristine fiber (above). Inelastic strain anisotropy as a function of demarcation energy (below).

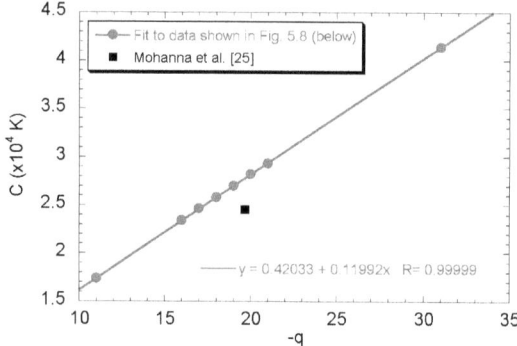

Figure 5.11: *For a fixed value of $b = 0.92$ in equation (5.10), good agreement between experimental data and the fit is found for several parameter pairs q, C. The two parameters correlate linearly.*

where $T_i = i\,\Delta t$ and the time interval Δt must be chosen small enough to guarantee convergence. The relaxational behavior characterized by equation (5.10) is illustrated as a solid line in Fig. 5.10 both for core stress (upper part) and strain anisotropy (lower part). The agreement with the experimentally observed core stress relaxation is satisfying, whereas the inelastic strain anisotropy only starts to relax at significantly higher demarcation energies/temperatures. We thus conclude that the relaxation of elastic strain introduced during fiber draw is thermally less stable than the inelastic contribution to strain anisotropy. However, it seems reasonable that the relaxation of inelastic strain is also governed by equations (5.9) and (5.10), only with a different parameter set (b, q, C). As we show later in section 5.2.2, the exponential factor for the relaxation of inelastic strain is found to be about $b = 0.92$. The two remaining parameters q and C can be found by fitting equation (5.10) to the experimental data. In Fig. 5.10 (below), good agreement is found for $q = -11$ and $C = 1.74 \times 10^4\,\text{K}$. However, fits of about the same quality can also be found with different pairs (q, C). As the experimental data are not sufficient to identify a particular fit as the best one, we illustrate in Fig. 5.11 all pairs (q, C), for which good agreement between fit and experiment has been found. Clearly, a linear correlation between the two parameters is observed. For comparison, the values found for stress relaxation by Mohanna *et al.* have also been included in the figure.

In Fig. 5.12, the changes in UV-induced axial core stress calculated according to equation (5.2) are illustrated as a function of refractive index demarcation energy. For $E_d < 3\,\text{eV}$, a linear correlation between refractive

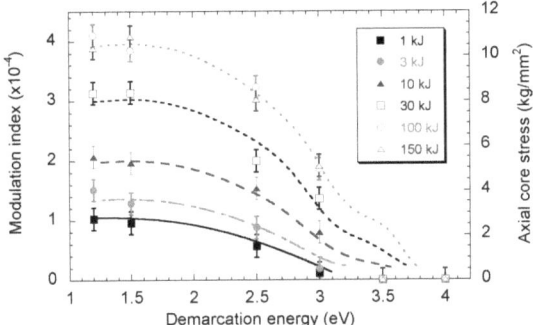

Figure 5.12: *Correlation of index modulation and UV-induced axial core stress changes for six different total doses.*

index and stress is found. However, the UV-induced axial core stress disappears for $E_d \geq 3\,\text{eV}$, although there remains still a significant amount of UV-induced refractive index. The refractive index change thus consists of at least two different contributions. The first contribution vanishes for $E_d < 3.5\,\text{eV}$ and is accompanied by a compaction of the core glass and a corresponding stress increase [13, 17]. For this contribution, two Gaussian energy distributions of defects have been identified in [22]. The other contribution withstands demarcation energies in excess of 3.5 eV. No difference in annealing behavior can be observed for the two highest doses of $100\,\text{kJ/cm}^2$ and $150\,\text{kJ/cm}^2$, respectively. By comparison of Fig. 5.12 with Fig. 5.10, it can be found that the UV-induced stresses anneal at considerably lower demarcation energies/temperatures than the drawing-induced stresses and inelastic strains.

Conclusion

Stress changes due to annealing have been observed for pristine as well as UV-irradiated germanium-doped fiber samples and correlated with refractive index annealing data. For refractive index demarcation energies below 3.5 eV, a linear correlation of about $0.9 \times 10^4 \text{kg/mm}^2$ between stress and index is found. For higher index demarcation energies, the stress is found to vanish, whereas remains of refractive index can still be observed. The UV-induced stresses anneal at lower demarcation energies in comparison to the drawing-induced stresses.

5.2 Phosphorus- and Nitrogen-doped fibers

5.2.1 Fibers under investigation

Phosphorus-doped fibers

Highly phosphorus-doped low-loss optical fibers are an attractive gain medium for Raman fiber amplifiers and lasers [26]. Highly efficient performance can be achieved by writing Bragg gratings as reflectors directly in the fiber used. Strong Bragg grating formation has been reported for hydrogenated phosphorus-doped fibers irradiated with ArF excimer radiation at 193 nm [27]. Photosensitivity in non-hydrogenated glass or under irradiation with 248 nm was found to be negligible [28]. As for Ge-doped fibers, the origins of photosensitivity in phosphorus-doped fibers have not yet been completely clarified. Some experiments suggest that densification is the dominating photosensitive mechanism [29].

Three phosphorus-doped single-mode fibers were drawn from the same preform manufactured by the modified chemical vapor deposition (MCVD) method at FORC. Inner cladding layers doped with 1 mol% P_2O_5 and 1 at% F and core layers doped with 12 mol% P_2O_5 have been deposited in a quartz tube. The refractive index profile of the preform was determined using a refractive near field method [30] and peaks at 1×10^{-2}. The fibers were drawn from the preform at temperatures of $T = 1860$, 1905 and 1940° C. The corresponding drawing tensions are $\tau = 225$, 115 and 70 g. In contrast to all other measurements presented within this thesis, the stress measurements on the three phosphorus-doped fibers have been performed using an older setup[6] with a laser scanning technique [31].

Nitrogen-doped fibers

Due to their stronger resistance to gamma-radiation as compared to germanium-doped fibers, nitrogen-doped fibers are a promising host fiber for Bragg gratings used as sensors in radiation environments [32]. Illumination of low-loss nitrogen-doped fibers results in an increase of the fibers refractive core index [33]. However, as for Ge-doped fibers, the origins of the change of index due to UV-irradiation have still not been completely identified.

The four fibers under investigation were drawn from the same preform at drawing tensions of 65, 125, 195 and 275 g, respectively. The corresponding drawing temperatures, at constant drawing speed, were 1940, 1905, 1880, and 1860° C. The preform has been synthesized by reduced-pressure surface plasma-chemical vapor deposition [34] incorporating a core doping of 1 at%

---

[6]Cabloptic S.A., 2016 Cortaillod, Switzerland.

nitrogen. The nitrogen-doped core layers were deposited in a pure silica supporting tube, which itself is embedded in several pure silica jacketing tubes. The refractive index difference was measured to peak at $(1.1\pm0.055)\times 10^{-2}$ for all fibers under investigation using a refractive near field method. The photo-elastic contribution to the refractive index difference was too small to be detected within the error.

5.2.2 Stress and inelastic strain changes with drawing tension

Phosphorus-doped fibers

The total axial stress distributions as a function of the radial position are shown in the upper part of Fig. 5.13 for the phosphorus-doped fibers. The inner cladding is under compressive stress and exhibits the stress minimum for all drawing temperatures. The core and the outer cladding tubes have about the same stress value for the fiber drawn with the lowest drawing tension of 60 g; however, for higher tensions, i.e. lower temperatures, the core stress decreases and turns compressive, whereas the cladding tube stresses increase. Two results are shown for the fiber drawn with a tension of 60 g. They were measured for samples coming from different positions along the fiber and thus reflect the variation of the axial stress profile along the fiber spool. This variation is caused by differences of drawing speed during the fiber fabrication process.

In the lower part of Fig. 5.13, the axial stress is plotted as a function of the drawing tension at three different radial positions of the fiber: in the center of the core at $r = 0\,\mu\text{m}$, in the inner cladding region (maximum of the compressive axial stress) at $r = 5\,\mu\text{m}$ and at the maximum of the tensile axial stress at $r = 62\,\mu\text{m}$. Within the error introduced by the variation of drawing speed during fabrication, a linear dependence between axial stress and drawing tension as predicted by equations (2.45) and (2.46) can be observed. The axial stress decreases linearly with drawing tension in the core and inner cladding regions, whereas it increases in the pure silica quartz tube. The corresponding slopes can be determined by fitting a linear function to the data. We find $\tilde{m}_{\text{Co}} = -42\,\text{mm}^{-2}$ in the core and $\tilde{m}_{\text{Cl}} = 8\,\text{mm}^{-2}$ in the cladding, respectively. For zero drawing tension, the remaining core stress is only thermally induced and can be found by extrapolation to be about $3.4\,\text{kg/mm}^2$. The corresponding free strain difference is found with equation (2.21) to be $\Delta\epsilon_f = -3.7\times 10^{-4}$.

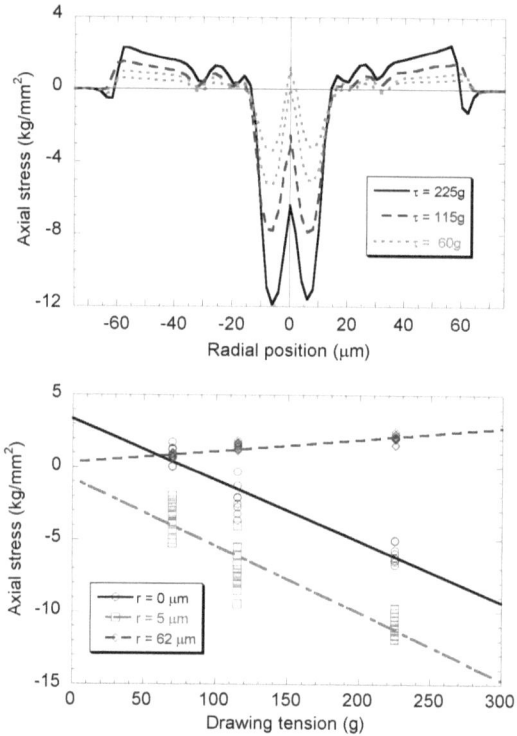

Figure 5.13: *Axial stress of phosphorus-doped fibers drawn from the same preform at different drawing temperature (above). The stresses are found to vary linearly with drawing tension (below).*

Nitrogen-doped fiber

The dependence of residual stress profile on drawing tension is illustrated for the nitrogen-doped fiber in the upper part of Fig. 5.14. As for the phosphorus-doped fiber, the core stress is found to increase with decreasing drawing tension and almost vanishes for the lowest tension of 65 g. Several cladding tubes can be identified that are all under tensile stress. As expected, the stress in the jacketing tubes increases with increasing drawing tension. The jacketing tubes are separated by regions with vanishing or even slightly compressive stress, that might result from OH-impurities [35].

The expected linear dependence between stress and drawing tension is depicted in the lower part of Fig. 5.14. The slopes for the core as well

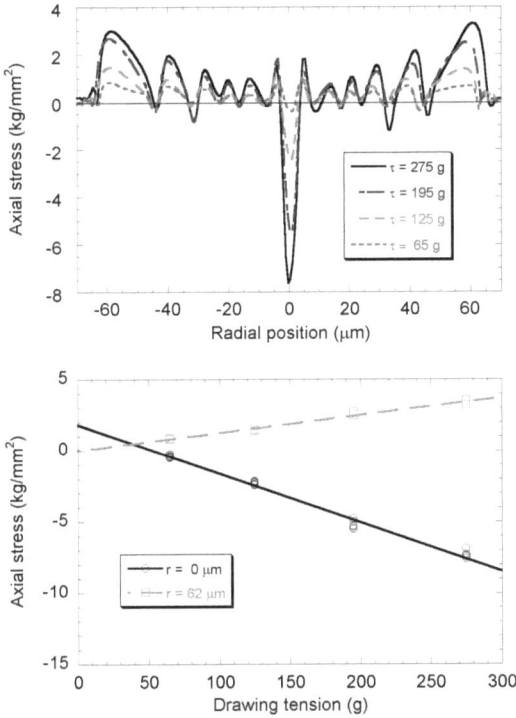

Figure 5.14: *Axial stress profiles for nitrogen-doped fibers as a function of drawing tension (above) and the corresponding linear dependence of stress on drawing tension in the core and outer cladding tube (below).*

as for the outer cladding tube, where the tensile stress is maximum, are again found by fitting a linear function and yield $\tilde{m}_{Co} = -34\,\text{mm}^{-2}$ and $\tilde{m}_{Cl} = 12\,\text{mm}^{-2}$, respectively. The axial core stress at zero drawing tension is found by extrapolation and yields $1.8\,\text{kg/mm}^2$, corresponding to a free strain difference of $\Delta\epsilon_f = -2 \times 10^{-4}$.

Thermally-induced stresses

If we assume constant thermal expansion coefficients, the temperature difference between the glass transition temperature T_S, where stresses start to develop, and room temperature can be deduced from equation (2.36). For the phosphorus-doped fiber, the thermal expansion coefficients of core and

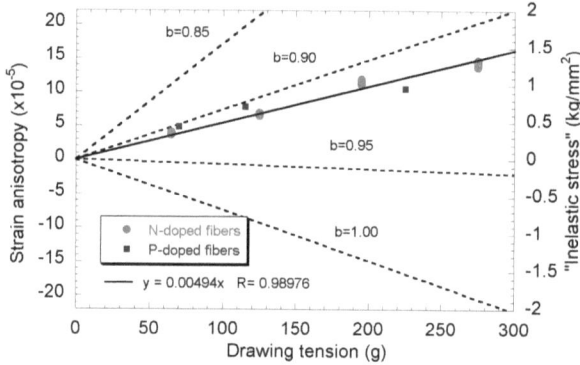

Figure 5.15: *Inelastic strain anisotropy of phosphorus- as well as nitrogen-doped fibers as a function of drawing tension. The two groups of fiber exhibit about the same dependence of strain anisotropy on drawing tension. The slope of the virtual "inelastic stress" with drawing tension is found by fitting and yields $\tilde{m} = 4.9\,\mathrm{mm}^{-2}$. The dotted lines indicate the expected dependence for different values of the exponential factor b of the Kohlrausch-Williams-Watt function, cf. equation (2.54).*

cladding are found from Fig. 2.1 to be $\alpha_1 = 20 \times 10^{-7}$ and $\alpha_1 = 5 \times 10^{-7}$, respectively. The free strain difference was found from the residual core stress at zero drawing tension and yields -3.7×10^{-4}. Thus, the difference between T_S and room temperature according to equation (2.36) is 250° C. Clearly, this temperature difference is smaller as expected; for slightly germanium-doped fibers, for example, thermal stresses start to develop already at about 1000° C [36]. A possible explanation might be a non-linear change of the thermal expansion coefficient with temperature. The values of the thermal expansion coefficient illustrated in Fig. 2.1 might thus only be valid in the lower temperature range. Unfortunately, reference [37] does not indicate the temperature range, for which the expansion has been determined.

For the nitrogen-doped fiber, no information about the thermal expansion coefficient could be found in the literature. As we neither know the temperature T_S where stress start to develop, it is difficult to draw any conclusions from equation (2.36).

Drawing-induced inelastic strain changes

All values for the slope of the axial stress as a function of drawing tension presented above comprise both a contribution due to real "elastic stress" and due to virtual "inelastic stress". To separate the two contributions, the "inelastic stress" has been calculated according to equation (4.13) for the two groups of fibers and is illustrated in Fig. 5.15. For both fibers, the strain anisotropy or "inelastic stress" respectively has almost the same dependence on drawing tension. As the strain anisotropy for silicate fibers is basically determined by the viscoelastic properties of the cladding material (cf. section 2.2.2), we conclude that the silicate tubes used for the manufacturing of the fiber must exhibit similar relaxational behavior. Thus, a single fitting can be applied to both fibers, resulting in a slope of $\tilde{m} = 4.9\,\text{mm}^{-2}$ for the dependence of "inelastic stress" on drawing tension.

In Fig. 5.15, the fitting curve is illustrated as a solid line as well as dotted curves obtained according to the theory depicted in section 2.2.2. Thus, the exponential factor b in equation (2.54) characterizing the relaxational behavior of the silica cladding can be adjusted to the data. Good agreement is found for a value of $b = 0.92$, which agrees well with values found in the literature for silica glasses [38].

Drawing-induced stress changes

According to equations (2.45) and (2.46), the drawing-induced stress profiles in core and cladding can be used to gather information about the viscosity and Young's modulus of the two regions. The cladding is in general made of pure silica, for which both constants are well established [39, 40]. Thus, equations (2.45) and (2.46) allow basically to assess their respective values in the core. However, the system of two equations can only be solved numerically. Analytical approximations can be found nevertheless, as Young's modulus only varies slightly between core and cladding [41, 42]. If it is assumed to be equal, each of the two equations (2.45) and (2.46) can essentially be used to express the core viscosity as a function of cladding viscosity, core and cladding area, as well as the slope of axial stress with drawing tension in core and cladding, respectively. As the axial stress change of the cladding with drawing tension varies as a function of radial position (cf. upper parts of Fig. 5.13 and 5.17), we use equation (2.45) to determine the core viscosity and get

$$\eta_1 = \frac{A_2 + \tilde{m}_{\text{Co}}(A_1 + A_2)}{A_2 - \tilde{m}_{\text{Co}}(A_1 + A_2)} \eta_2. \tag{5.11}$$

where the slope of axial core stress with drawing tension $\tilde{m}_{\text{Co}}$ is given in mm^{-2}. Another reason for using the slope in the core and not in the cladding is that the core as the material with the higher viscosity should be less affected by inelastic strains.

For the phosphorus-doped fiber, we found $\tilde{m}_{\text{Co}} = -42\,\text{mm}^{-2}$ in Fig. 5.13 below. The corresponding dependence of core viscosity on cladding viscosity

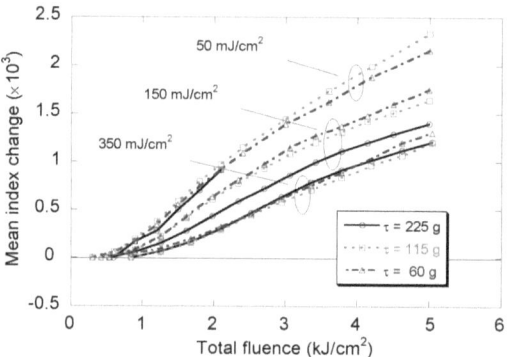

Figure 5.16: *UV-induced mean index changes as a function of drawing tension and pulse fluence for the phosphorus-doped fiber.*

yields $\eta_1 = 0.48\,\eta_2$ from equation (5.11), where a core radius of 3 µm has been assumed. If we subtract the slope of $\tilde{m} = 4.9\,\text{mm}^{-2}$ due to the virtual "inelastic stress" established in the previous section, we get a core slope of $\tilde{m}_{\text{Co}} = -37.5\,\text{mm}^{-2}$ and a corresponding core viscosity of $\eta_1 = 0.54\,\eta_2$. Unfortunately, no reference about the dependence of viscosity on dopant concentration is available for phosphorus-doped silica.

For the nitrogen-doped fiber, the slope of $\tilde{m}_{\text{Co}} = -34\,\text{mm}^{-2}$ corresponds to a core viscosity of $\eta_1 = 0.58\,\eta_2$, where a core radius of 2 µm has been assumed in equation (5.11). If we subtract the slope due to "inelastic stress", we get $\eta_1 = 0.64\,\eta_2$. Again, no comparison with already published data is possible.

5.2.3 UV-induced stress changes

Phosphorus-doped fibers

Prior to UV-exposure, the phosphorus-doped fibers were soaked in hydrogen for 48 h at a pressure of 100 bar and a temperature of 100° C. The resulting hydrogen-concentration in the fiber is estimated to be 5.5×10^3 ppm [43]. Stress measurements have been performed on the fibers after they had been removed from the hydrogen-chamber. However, if stress changes occurred during hydrogen loading, they must have been smaller than the error introduced by the variation in drawing force (up to ±35%).

In all fibers, Bragg gratings have been inscribed directly after the fibers had been removed from the hydrogen-chamber. The gratings were written with an excimer-laser at 193 nm using the phase mask technique described in section 3.4.1. The phase mask period was 1.018 µm, and the grating length 8 mm. For all gratings, the total fluence was 5 kJ/cm^2 at a pulse duration of 20 ns and a pulse repetition rate of 20 Hz. Three different pulse fluences (50, 150 and 350 mJ/cm^2) were used. Thus, in each of the three fibers three gratings have been written. The grating spectra have been recorded using a spectrum analyzer with a resolution of 0.1 nm. The mean index-change was determined via the Bragg wavelength shift and is illustrated in Fig. 5.16 as a function of total dose for the nine gratings under investigation. Drawing conditions are found not to alter the index change significantly, as has already been shown for hydrogenated Ge-doped fibers [44]. Decreasing pulse fluence, however, was found to increase the refractive index change for all fibers.

In Fig. 5.17, axial stress profiles are shown before and after grating-inscription for gratings written with a pulse fluence of 350 mJ/cm^2 in the three fibers. The total dose is 5 kJ/cm^2 for all gratings illustrated in Fig. 5.17. The stress profiles of the non-irradiated fibers were recorded directly next to the grating, so the influence of stress changes over the length of the fiber spool can be neglected. For all fibers, the core stress is found to increase by about 2 kg/mm^2 due to UV-illumination. The increase in axial core stress does not depend on the initial stress state, compressive or tensile. A similar result has already been reported earlier for Ge-doped non-hydrogenated fibers [13].

The change of axial stress for the fiber drawn with a tension of 70 g is shown in Fig. 5.18 for different pulse fluences. The total dose for all gratings is again 5 kJ/cm^2. No correlation between pulse fluence and axial stress change can be found within the accuracy of measurement. This observation has also been made for the fibers drawn with 115 and 225 g.

Enhanced index growth is found for lower pulse fluences. The difference in refractive index is about 0.5×10^{-3} (see Fig. 5.16), which corresponds to a stress change of about 7.5 kg/mm^2 [17]. This large difference in stress cannot be confirmed by stress measurements (Fig. 5.18). The difference in index growth for different pulse fluences thus does not seem to be a result of the densification process.

The increase in axial tensile stress was found to be about 2 kg/mm^2 for all gratings under observation. The corresponding amount of core densification can be found with equation (2.40) and yields $\Delta\rho/\rho = -\Delta V/V = 0.066\,\%$. The corresponding net index contribution (equation (3.12)) of 0.9×10^{-4} is composed of an inelastic part of 2.3×10^{-4} and a stress-induced index response of -1.4×10^{-4}.

All gratings were written with the same total dose, but their refractive index difference differed by more than a factor of two as a function of pulse fluence. Thus, a linear relationship between axial stress and index change as found for germanium-doped fibers does not seem to apply for phosphorus

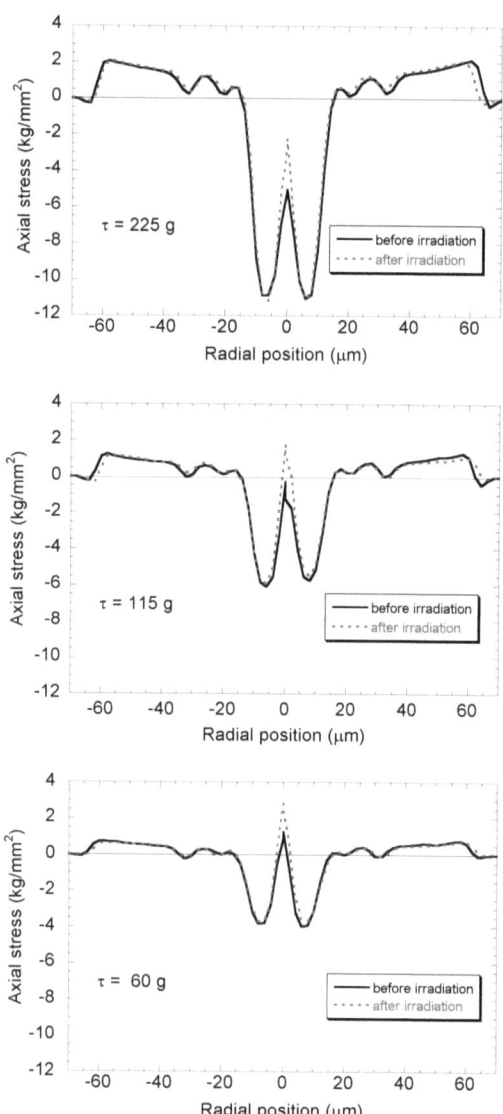

Figure 5.17: Axial stress changes for the three phosphorus-doped fibers due to UV-illumination with a pulse fluence of $350\,\mathrm{mJ/cm^2}$. The total dose is $5\,\mathrm{kJ/cm^2}$ for all gratings. The axial core stress is found to increase for all fibers.

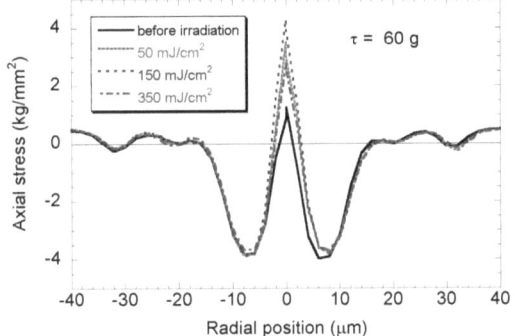

Figure 5.18: Axial stress change at different pulse fluences for the phosphorus-doped fiber drawn with 60 g. The total dose is 5 kJ/cm² for all gratings. Within the accuracy of stress measurement, no correlation between pulse fluence and stress change is found.

fibers. However, by using a mean index change value of $\Delta n = (1.5 \pm 0.5) \times 10^{-3}$, we can determine an approximate value of stress change with index which is $\Delta\sigma_{zz}/\Delta n \approx (0.13\pm 0.04)\times 10^4$ kg/mm². The amount of stress change for a given index change is thus almost one order of magnitude smaller for phosphorus fibers than for germanium-doped fibers, indicating a considerably smaller contribution of densification to the overall index change.

Nitrogen-doped fibers

To investigate the influence of UV-illumination on residual stress changes in nitrogen-doped fibers, the four fibers under investigation were side exposed to an ArF excimer laser beam ($\lambda = 193$ nm). The pulse fluence was adjusted to (100 ± 20) mJ/cm². Two series of fiber samples were fabricated, differing in pulse number and thus total fluence. The first series was exposed to 12000, the second one to 24000 pulses. Repetition rate was 20 Hz. The resulting total fluences were 1.2 kJ/cm² and 2.4 kJ/cm². As no gratings were written into the fiber in this experiment, the refractive index change could not be determined online. For comparable irradiation conditions, however, index changes of about 1×10^{-4} and 2.2×10^{-4} have been reported recently in nitrogen-doped fibers [33].

As shown in Fig. 5.19 for the fiber drawn with a tension of 275 g, the core stress is increased by 0.8 kg/mm² after irradiation with 12000 pulses and by further 0.7 kg/mm² after irradiation with 24000 pulses. No photo-induced asymmetry is found in the fiber core. The increase in core stress does

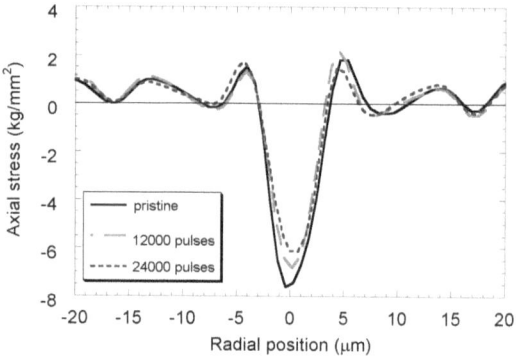

Figure 5.19: *Axial core stress changes with total dose for the nitrogen-doped fiber drawn with* 275 g. *UV-irradiation enhances the compressive core stress.*

not depend on drawing tension and thus initial stress state, as illustrated in Fig. 5.20. The increase in stress due to UV-illumination is about 0.6×10^{-3} kg/mm^2/pulse for all four drawing tensions.

The knowledge of both axial stress increase and refractive index change again allows to estimate the core densification and its contribution to the overall index change. The maximum stress change found for an irradiation with 24000 pulses is about 1.5 kg/mm^2 for all fibers under investigation. The corresponding core densification is about 0.055 %. The resulting densification induced index change is 1.7×10^{-4} according to equation (3.8). The photoelastic response compensates for an index change of 1×10^{-4}, so that the net index change given by equation (3.12) becomes $\Delta n_{\perp}^{tot} = 0.7 \times 10^{-4}$.

The refractive index change of nitrogen-doped fibers not only depends on the total dose, but also on the pulse fluence [33]. Here, the axial stress change correlates linearly with the experimental index change reported in [33]. The amount of stress change with total index change is about $\Delta \sigma_{zz}/\Delta n \approx 0.75 \times 10^4$ kg/mm^2. The densification induced index contribution is thus larger than in hydrogenated phosphorus-doped fibers, but slightly smaller than in germanium-doped fibers.

5.3 Summary of results and discussion

The results presented in this chapter allow to summarize the contribution of core densification, manifesting itself in core stress increase, to the total index change for optical fibers with different core dopants. For all fibers investigated so far, an increase in core stress occurred after irradiation with

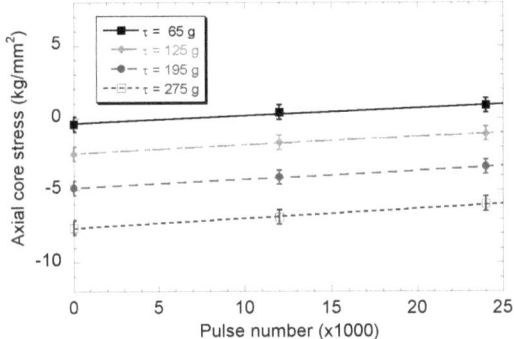

Figure 5.20: *Axial stress dependence of the nitrogen-doped fiber on drawing tension and UV-irradiation. The stress decreases linearly with increasing drawing tension. UV-irradiation enhances the compressive core stress.*

a UV- or femtosecond laser. Thus, a change in refractive index seems always to be accompanied by an axial core stress increase.

According to the theory developed in section 3.3 of the third chapter, axial stress changes can be interpreted as an indication of core densification. The core densification, in turn, changes the stress profile of the fiber. The total amount of densification-induced refractive index change thus comprises two contributions: an inelastic, positive one due to compaction, and an elastic, negative one due to densification-induced stress changes. The elastic index response compensates for almost 60 % of the inelastic one, i.e. the compaction efficiency only yields about 40 %.

As the axial stress change is linearly related to both densification and net densification-induced index change, its correlation with the overall refractive index change characterizes the amount of densification-induced index change to overall index change due to both densification and color-center formation. If the overall index change was only densification-induced, than its correlation with stress would be given by equation (3.13) and yield $\Delta\sigma_{zz}/\Delta n = 2.2 \times 10^4 \, \text{kg/mm}^2$.

Fonjallaz *et al.* found a linear dependency between axial stress and overall refractive index increase of $\Delta\sigma_{zz}/\Delta n = (1.3 \pm 0.3) \times 10^4 \, \text{kg/mm}^2$ for germanium-doped fibers, independently of dopant level, total dose, and pulse fluence. The pulsed irradiation source was a frequency-doubled excimer-pumped dye laser emitting at 240 nm. Raine *et al.* reported results for 193 nm and 248 nm excimer-irradiation in [45] that are in broad agreement with Fonjallaz' results. In section 5.1.2 of this chapter, we find a linear re-

Dopant	Source	$\Delta\sigma/\Delta n$ ($\times 10^4$ kg/mm^2)	$\Delta n_\perp^{tot}/\Delta n$ (%)	Ref.
–	–	2.2	100	eq. (3.13)
GeO$_2$ (3 to 18 mol%)	248 nm pulsed	1.3 ± 0.3	60 ± 14	[13]
GeO$_2$ (3 mol%)	810 nm fs-laser	1.55 ± 0.15	69 ± 7	this work
GeO$_2$ (14 mol%)	244 nm cw-laser	0.9 ± 0.1	41 ± 5	this work
P$_2$O$_5$ (14 mol%) & H$_2$	193 nm pulsed	0.13 ± 0.04	6 ± 1.7	this work
N (1 at%)	193 nm pulsed	0.75 ± 0.5	34 ± 22	this work

Table 5.1: *Comparison of stress changes with total index change and corresponding densification-induced index contribution for fibers with different core dopants and irradiation conditions.*

lationship between stress and index change for cw-irradiation at 244 nm of 0.9×10^4 kg/mm^2, which is only slightly smaller than the value reported by Fonjallaz et al. Furthermore, the correlation of stress with index changes for fs-laser induced photosensitivity yields 1.5×10^4 kg/mm^2, which is also in agreement with Fonjallaz' results. The linear relationship between stress and index change reported by Fonjallaz thus seems to apply for a variety of irradiation conditions and germanium-concentrations. The corresponding ratio of densification-induced to overall index change is about $(60 \pm 12\,\%)$.

Radiation-induced core stress changes in hydrogen-loaded phosphorus-doped fibers have been reported in this thesis for the first time. The stress changes only depend on total fluence, and not on the initial drawing-induced stress profile or pulse fluence. The radiation-induced index change, however, was found to increase considerably for decreasing pulse fluences. The ratio between overall index change and axial stress change thus also depends on the specific irradiation conditions in the case of phosphorus-doped fibers. To estimate nevertheless the order of magnitude of densification-induced index change with respect to overall index change, we correlate the stress increase of 2 kg/mm^2 with the index increase of $(1.6 \pm 0.4) \times 10^{-3}$ found by averaging over the three different pulse fluences. We get $\Delta\sigma_{zz}/\Delta n = (0.13 \pm 0.04) \times 10^4$ kg/mm^2, a value that is about one order of magnitude smaller than for germanium-doped fibers. The corresponding contribution of densification-induced to overall index change is consequently also about one order of magnitude smaller and yields $(6 \pm 1.7)\,\%$.

In addition, we report for the first time about irradiation-induced stress changes in nitrogen-doped silica fibers. For the irradiation condition used within this experiment, both axial core stress as well as refractive index increase linearly with total dose. This again allows to estimate a ratio between stress and index change. We find $\Delta\sigma_{zz}/\Delta n = (0.75 \pm 0.5) \times 10^4\,\text{kg}/\text{mm}^2$, where we used a relative error of 50 % for the refractive index change. The densification-induced contribution to the overall index change thus reads $(34 \pm 22)\%$ As the refractive index change for nitrogen-doped fibers depends strongly on irradiation conditions [33], this value might only be appropriate for the small pulse fluence used in our experiment.

Our results about stress and overall index correlations and the resulting densification-induced index contributions are compiled in Table 5.1. Clearly, the densification induced index change is smallest for the hydrogenated phosphorus-doped fiber. Hydrogen-loading might thus especially favor color center formation and respective index changes, whereas its effect on radiation-induced densification is small.

Bibliography

[1] F. Dürr, H. G. Limberger, R. P. Salathé, F. Hindle, M. Douay, E. Fertein, and C. Przygodzki. Tomographic measurement of femtosecond-laser induced stress changes in optical fibers. *Applied Physics Letters*, 84(24):4983–4985, 2004.

[2] F. Dürr, H. G. Limberger, R. P. Salathé, S. A. Vasiliev, O. I. Medvedkov, A. S. Bozhkov, and E. M. Dianov. Annealing-induced stress changes in UV-irradiated germanium-doped fibers. In *BGPP 2005 Technical Digest*. Optical Society of America, 2005. accepted as poster presentation.

[3] F. Dürr, H. G. Limberger, R. P. Salathé, F. Cochet, A. A. Rybaltovsky, Y. V. Larionov, S. L. Semjonov, and E. M. Dianov. UV-induced stress changes in phosphorus-doped fibers drawn at different drawing tensions. In *BGPP 2003 Technical Digest*. Optical Society of America, 2003. paper BThA2.

[4] N. Bloembergen. Laser-induced electric breakdown in solids. *IEEE Journal of Quantum Electronics*, 10(3):375–386, 1974.

[5] C. B. Schaffer, A. Brodeur, and E. Mazur. Laser-induced breakdown and damage in bulk transparent materials induced by tightly focused femtosecond laser pulses. *Measurement Science & Technology*, 12(3):1784–1794, 2001.

[6] N. T. Nguyen, A. Saliminia, W. Liu, S. L. Chin, and R. Vallée. Optical breakdown versus filamentation in fused silica by use of femtosecond infrared laser pulses. *Optics Letters*, 28(17):1591–1593, 2003.

[7] K. M. Davis, K. Miura, N. Sugimoto, and K. Hirao. Writing waveguides in glass with a femtosecond laser. *Optics Letters*, 21(21):1729–1731, 1996.

[8] A. M. Streltsov and N. F. Borrelli. Study of femtosecond-laser-written waveguides in glasses. *Journal of the Optical Society of America B*, 19(10):2496–2504, 2002.

[9] Y. Kondo, K. Nouchi, T. Mitsuyu, M. Watanabe, P. G. Kazansky, and K. Hirao. Fabrication of long-period fiber gratings by focused irradiation of infrared femtosecond laser pulses. *Optics Letters*, 24(10):646–648, 1999.

[10] S. J. Mihailov, C. W. Smelser, P. Lu, R. B. Walker, D. Grobnic, H. Ding, G. Henderson, and J. Unruh. Fiber Bragg gratings made with a phase mask and 800 nm femtosecond irradiation. *Optics Letters*, 28(12):995–997, 2003.

[11] J. W. Chan, T. Huser, S. Risbud, and D. M. Krol. Structural changes in fused silica after exposure to focused femtosecond laser pulses. *Optics Letters*, 26(21):1726–1728, 2001.

[12] E. M. Dianov, V. G. Plotnichenko, V. V. Koltashev, N. Pyrkov Yu, N. H. Ky, H. G. Limberger, and R. P. Salathé. UV-irradiation-induced structural transformation of germanosilicate glass fiber. *Optics Letters*, 22(23):1754–1756, 1997.

[13] P. Y. Fonjallaz, H. G. Limberger, R. P. Salathe, F. Cochet, and B. Leuenberger. Tension increase correlated to refractive-index change in fibers containing UV-written Bragg gratings. *Optics Letters*, 20(11):1346–1348, 1995.

[14] A. Brodeur and S. L. Chin. Ultrafast white-light continuum generation and self-focusing in transparent condensed media. *Journal of the Optical Society of America B*, 16(4):637–650, 1999.

[15] F. Hindle, E. Fertein, C. Przygodzki, F. Dürr, L. Paccou, R. Bocquet, P. Niay, H. G. Limberger, and M. Douay. Inscription of long-period gratings in pure silica and germano-silicate fiber cores by femtosecond laser irradiation. *IEEE Photonics Technology Letters*, 16(8):1861–1863, 2004.

[16] Y. Park, U. C. Paek, and D. Y. Kim. Complete determination of the stress tensor of a polarization-maintaining fiber by photoelastic tomography. *Optics Letters*, 27(14):1217–1219, 2002.

[17] H. G. Limberger, P. Y. Fonjallaz, R. P. Salathe, and F. Cochet. Compaction- and photoelastic-induced index changes in fiber Bragg gratings. *Applied Physics Letters*, 68(22):3069–3071, 1996.

[18] J. Nishii. Permanent index changes in Ge-SiO_2 glasses by excimer laser irradiation. *Materials Science & Engineering B*, 54:1–10, 1998.

[19] H. G. Limberger, P. Y. Fonjallaz, and R. P. Salathe. Spectral characterisation of photoinduced high efficiency Bragg gratings in standard telecommunication fibres. *Electronics Letters*, 29(1):47–49, 1993.

[20] T. Erdogan, V. Mizrahi, P. J. Lemaire, and D. Monroe. Decay of ultraviolet-induced fiber Bragg gratings. *Journal of Applied Physics*, 76(1):73–80, 1994.

[21] J. Rathje, M. Kristensen, and J. E. Pedersen. Continuous anneal method for characterizing the thermal stability of ultraviolet Bragg gratings. *Journal of Applied Physics*, 88(2):1050–1055, 2000.

[22] S. A. Vasilev, O. I. Medvedkov, A. S. Bozhkov, and E. M. Dianov. Annealing of UV-induced fiber gratings written in Ge-doped fibers: investigation of dose and strain effects. In *BGPP 2003 Technical Digest*. Optical Society of America, 2003. paper MD31.

[23] M. J. LuValle, L. R. Copeland, S. Kannan, J. B. Judkins, and P. J. Lemaire. A strategy for extrapolation in accelerated testing. *Bell Labs Technical Journal*, July-September 1998.

[24] W. Hermann, M. Hutjens, and D. U. Wiechert. Stress in optical waveguides. III. Stress induced index change. *Applied Optics*, 28(11):1980–1983, 1989.

[25] Y. Mohanna, J. M. Saugrain, J. C. Rousseau, and P. Ledoux. Relaxation of internal stresses in optical fibers. *Journal of Lightwave Technology*, 8(12):1799–1801, 1990.

[26] E. M. Dianov, I. A. Bufetov, M. M. Bubnov, A. V. Shubid, S. A. Vasiliev, O. I. Medvedkov, S. L. Semjonov, M. V. Grekov, V. M. Paramonov, A. N. Gur'yanov, V. F. Khopin, D. Varelas, A. Iocco, D. Costantini, H. G. Limberger, and R. P. Salathe. CW highly efficient 1.24 mu m Raman laser based on low-loss phosphosilicate fiber. In *OFC'99 Technical Digest*. Optical Society of America, 1999. paper PD25.

[27] T. A. Strasser, A. E. White, M. F. Yan, P. J. Lemaire, and T. Erdogan. Strong Bragg phase gratings in phosphorus-doped fiber induced by ArF excimer radiation. In *OFC'95 Technical Digest*, pages 159–160. Optical Society of America, 1995.

[28] B. Malo, J. Albert, F. Bilodeau, T. Kitagawa, D. C. Johnson, K. O. Hill, K. Hattori, Y. Hibino, and S. Gujrathi. Photosensitivity in phosphorus-doped silica glass and optical waveguides. *Applied Physics Letters*, 65(4):394–396, 1994.

[29] H. Hosono, K. Kawamura, and M. Hirano. Defect formation in SiO_2:P_2O_5 glasses by Excimer laser irradiation: effects of hydrogen loading. In *BGPP 2001 Technical Digest*. Optical Society of America, 2001. paper BThA2.

[30] K. W. Raine, J. G. N. Baines, and D. E. Putland. Refractive index profiling-state of the art. *Journal of Lightwave Technology*, 7(8):1162–1169, 1989.

[31] P. L. Chu and T. Whitbread. Measurement of stresses in optical fiber and preform. *Applied Optics*, 21(23):4241–4245, 1982.

[32] E. M. Dianov, K. M. Golant, R. R. Khrapko, and A. L. Tomashuk. Nitrogen doped silica core fibres: a new type of radiation-resistant fibre. *Electronics Letters*, 31(17):1490–1491, 1995.

[33] E. M. Dianov, K. M. Golant, R. R. Khrapko, A. S. Kurkov, B. Leconte, M. Douay, P. Bernage, and P. Niay. Grating formation in a germanium free silicon oxynitride fibre. *Electronics Letters*, 33(3):236–238, 1997.

[34] E. M. Dianov, K. M. Golant, R. R. Khrapko, A. S. Kurkov, and A. L. Tomashuk. Low-hydrogen silicon oxynitride optical fibers prepared by SPCVD. *IEEE Journal of Lightwave Technology*, 13(7):1471–1474, 1995.

[35] B. H. Kim, Y. Park, D. Y. Kim, U. C. Paek, and W. T. Han. Observation and analysis of residual stress development resulting from OH impurity in optical fibers. *Optics Letters*, 27(10):806–808, 2002.

[36] Y. Y. Huang, A. Sarkar, and P. C. Schultz. Relationship between composition, density and refractive index for germania silica glasses. *Journal of Non-Crystalline Solids*, 27(1):29–37, 1978.

[37] S. R. Nagel. Silica-based glass optical fiber properties and fabrication methods. *Proceedings of the SPIE*, 1085:56–76, 1990.

[38] R. Böhmer, K. L. Ngai, C. A. Angell, and D. J. Plazek. Nonexponential relaxations in strong and fragile glass formers. *Journal of Chemical Physics*, 99(5):4201–4209, 1993.

[39] W. Primak and D. Post. Photoelastic constants of vitreous silica and its elastic coefficient of refractive index. *Journal of Applied Physics*, 30(5):779–788, 1959.

[40] R. H. Doremus. Viscosity of silica. *Journal of Applied Physics*, 92(12):7619–7629, 2002.

[41] M. Yamane and T. Sakaino. Calculation of Young's modulus of glass from its chemical composition and density. *Glass Technology*, 15(5):134–136, 1974.

[42] P. K. Bachmann, W. Hermann, H. Wehr, and D. U. Wiechert. Stress in optical waveguides. II. Fibers. *Applied Optics*, 26(7):1175–1182, 1987.

[43] P. J. Lemaire. Reliability of optical fibers exposed to hydrogen: prediction of long-term loss increases. *Optical Engineering*, 30(6):780–789, 1991.

[44] I. Riant and B. Poumellec. Influence of fiber drawing tension on photosensitivity in hydrogenated and nonhydrogenated fibers. In *OFC'98 Technical Digest*. Optical Society of America, 1998. paper TuA1.

[45] K. W. Raine, R. Feced, S. E. Kanellopoulos, and V. A. Handerek. Measurement of axial stress at high spatial resolution in ultraviolet-exposed fibers. *Applied Optics*, 38(7):1086–1095, 1999.

Chapter 6

Fiber characterization using Atomic Force Microscopy

A short overview about the possibilities to assess optical fibers properties by scanning their etched end-faces with an Atomic Force Microscope (AFM) has been given in section 3.4.2. Here, we report on etch rate changes in fibers due to hydrogen-loading, UV-irradiation and fiber drawing properties[1]. The observed etch rate modifications suggest the potential of the AFM-technique not only to determine dopant concentrations and corresponding refractive index profiles, but also to gather information about the formation of color-centers during fiber fabrication and their redistribution due to UV-irradiation.

6.1 Phosphorus-doped fibers

6.1.1 Experiment

The fiber investigated within this study is the phosphorus-doped fiber drawn at 225 g which has already been described in section 5.2.1 of the preceding chapter. There, its axial core stress was found to be compressive with a value of $-5\,\text{kg/mm}^2$. For the AFM-study, the fiber was hydrogen loaded for about two weeks at a pressure of 110 bar at room temperature. Two pieces of hydrogenated fiber were subsequently exposed to 193 nm light from an excimer-laser under different radiation conditions, both resulting in a total fluence of $1\,\text{kJ/cm}^2$. For the first fiber, the pulse fluence was $50\,\text{mJ/cm}^2$ at a repetition rate of 1 Hz, resulting in an exposure time of about 330 minutes. The second fiber was illuminated with a pulse fluence of $350\,\text{mJ/cm}^2$ at 20 Hz repetition rate and a corresponding exposure time of 142 seconds.

---

[1]results published in [1], [2]

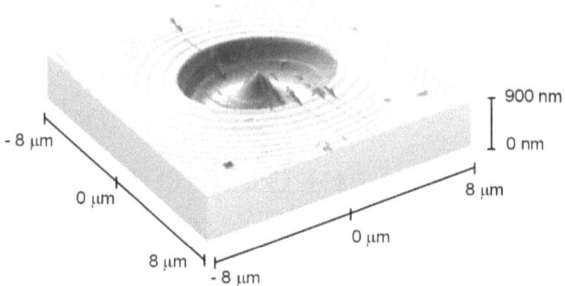

Figure 6.1: AFM image of phosphorus-doped fiber after etching in hydrofluoric acid for 180 seconds. The core is observed as a crater since it is etched faster than the concentric inner cladding SiO_2 layers.

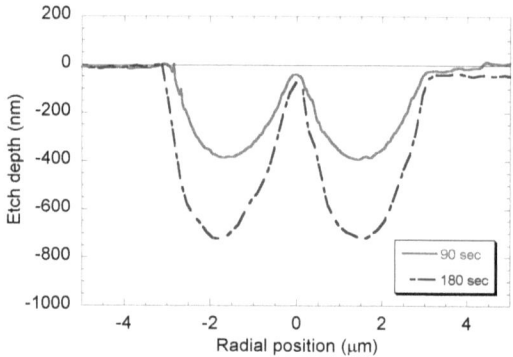

Figure 6.2: Etch dynamics of the pristine phosphorus-doped fiber core for two different etch times. The difference in diameter reflects the core width error due to side etching [3].

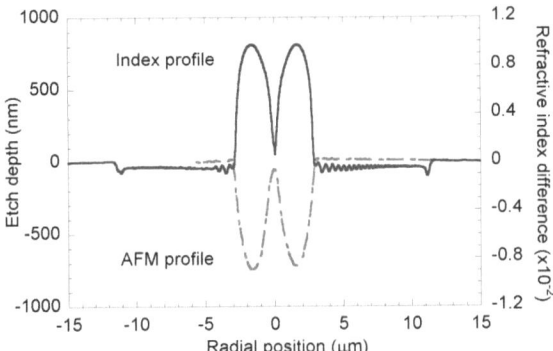

Figure 6.3: Comparison of the preforms refractive index profile with the etch profile of the pristine fiber. The preform diameter has been downscaled for comparison.

For AFM-observation, the samples were cleaved and etched in a 5 % HF solution for 90 and 180 seconds, respectively. Between hydrogen loading and etching, the samples were stored for about one week at room temperature to ensure that the interstitial hydrogen was out-diffused from the fiber. The AFM (TopoMetrix Explorer) was operated in contact mode in air with a standard V-shaped silicon nitride cantilever.

6.1.2 Etch dynamics of the pristine fiber

In Fig. 6.1, the topography of the pristine fiber after an etching time of 180 seconds is illustrated. The highly phosphorus-doped core layers are etched faster than the surrounding pure silica inner cladding layers. The dip in the center of the core is caused by dopant out diffusion during collapsing of the preform tube. The core is found to be elliptical; the diameters (FWHM) of the two main axes are 4.75 µm and 5.4 µm, respectively.

The etch profile of the pristine fiber is illustrated for two different etch times in Fig. 6.2. The profiles have been aligned along the major axis of the elliptical core for comparison. The etch depth doubles for an increase in etch time from 90 to 180 seconds, thus confirming the linear dependence of etch depth on etch time already reported for germanium-doped fibers [4]. The increase in diameter from 5.1 µm (90 seconds) to 5.4 µm (180 seconds) between the two profiles reflects the core width error due to side etching [3].

Fig. 6.3 illustrates the etch profile of the pristine fiber etched for 180 seconds together with the refractive index profile of the fiber preform. For comparison, the preform diameter has been downscaled to the diameter of the fiber. As the inner cladding of the fiber is co-doped with 1 at% F, the outer pure silica tubes have been chosen as reference level for the refractive index profile. If we assume that the phosphorus content in the fiber depends linearly on refractive index difference according to $[P_2O_5](\text{mol}\%) = \Delta n/(0.88 \times 10^{-3})$ [5], the refractive index profile can be converted into a dopant profile.

Pace et al. reported a power-law dependence between dopant concentration and etch rate for germanium-doped fibers in [3]. To obtain similar information for phosphorus-doped fibers, the etch profile and the preform profile shown in Fig. 6.3 have been correlated within the doped core region. The etch rate was determined by dividing the absolute value of the etch depth through the corresponding etch time. The correlation curve between index/phosphorus-concentration and etch rate is shown in Fig. 6.4. It can be fitted to the power-law function

$$\frac{dx}{dt} = k(\Delta n)^n \qquad (6.1)$$

where dx/dt is the etch rate, k the reaction rate constant, Δn the refractive index and n the reaction order. The reaction constants thus obtained are $k = (4.31 \pm 0.04)$ nm/s and $n = 2.15 \pm 0.06$, and the corresponding fitting curve is illustrated in Fig. 6.4. The reaction order of 2.15 is slightly higher than the corresponding value of about 1.7 for germanium-doped silica [3].

6.1.3 Hydrogen- and UV-induced etch rate changes

In a next step, AFM-profiles obtained for hydrogenated as well as irradiated fiber have been compared with the profiles from the pristine fiber. In Fig. 6.5, the maximum etch depth occurring at a radial position of about $\pm 1.5\,\mu\text{m}$ is plotted as a function of etch time for all samples under investigation. A linear dependence of etching depth on time is confirmed for the pristine, hydrogen-loaded, and UV-irradiated etched fiber cores. The slope of the linear fit defines the etch rate of the corresponding sample. The etch rate is about 4 nm/s for the pristine fiber and is reduced by almost 30 % due to hydrogen loading. In comparison to the etch profile of the hydrogenated as well as the pristine fiber, the etch depth of the core region is increased by UV-irradiation for both pulse fluences. However, no significant dependence of etch rate on pulse fluence can be observed. The etch rate of the irradiated samples has increased by about 27 % as compared to the pristine sample and by about 75 % as compared to the hydrogenated sample.

All profiles were found to scale linearly with etch time, as it was shown for the pristine fiber in Fig. 6.2. Furthermore, the profile of the pristine fiber was found to scale almost linearly with the profiles obtained after hydrogenation

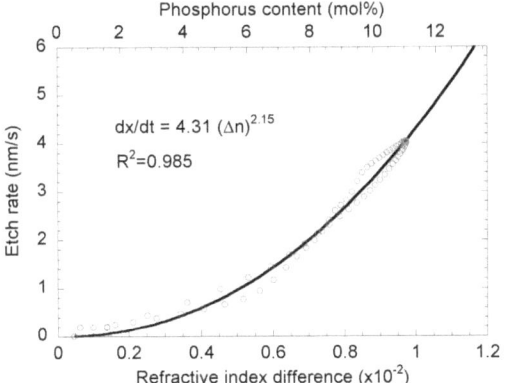

Figure 6.4: *Correlation of the etch rate with refractive index difference and phosphorus content, respectively. As for germanium-doped fibers [3], a power-law dependence of etch rate on refractive index is found.*

and irradiation. Particularly, no preferential etching at the core-cladding interface can be found in contrast to already published results by Canning et al. [6]. As the shape of the profile does not change for all samples, only the reaction constant, not the reaction order of the process is modified by fiber treatment.

6.1.4 Discussion of results

The refractive index profile of optical fibers does not only depend on the dopant concentration in the core, but also on the difference in color-center concentrations [7] and stress [8] between core and cladding. Both properties are altered during the fiber drawing process as well as by UV-irradiation [9]. Also hydrogen loading was found to generate new defects in phosphorus-doped glass samples [7] and to modify residual stresses in Sn- and B- codoped germanosilicate fibers [10]. In the previous section, we assumed a linear dependence between refractive index and dopant profile of $\Delta n/[P2O5] = 0.88 \times 10^{-3}$, found in reference [5] for the fiber under investigation. In the same reference, several values for the ratio of refractive index and dopant concentration reported by different groups have been compared. The results were found to vary considerably within 0.83 to 1.3 $\times 10^{-3}$, depending on the sample processing conditions. The scattering of data even for equally doped samples has been explained primarily by the occurrence of stress, modifying

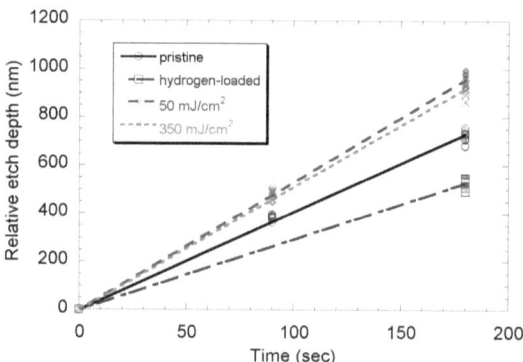

Figure 6.5: Relative etch depth as function of time for the etch maxima. For all samples under investigation, the etch maxima depend linearly on time.

the index through the photo-elastic effect. As the variance in etch rates found within this study is comparable to the variance in molar refractivity reported in [5], the difference in etch rate might also be caused by stress or defects introduced into the fiber core during fabrication. In the following discussion, we correlate our etch rate results with reported changes in refractive index, stress and color center concentration introduced by hydrogen-loading and subsequent UV-irradiation. In this way, we are able to isolate the parameter determining the differences in etch rate.

Correlation of etch rate changes with refractive index changes

For the fiber under investigation, the total index change between core and inner cladding is $\Delta n = 1 \times 10^{-2}$. For phosphorus-doped glass samples, the reported changes in UV absorption due to hydrogen loading are small [7]. The corresponding index changes, estimated using the Kramers-Kronig relationship, are on the order of some ppm. Mean Bragg-grating index changes on the order of several 10^{-4} have been found after irradiating the fiber with an ArF-laser to a total fluence of $1\,\text{kJ/cm}^2$ [11]. If the reaction rate was linearly related to refractive index, the increase in reaction rate after UV-irradiation should almost be one order of magnitude smaller than the 27 % reported in the previous section. Furthermore, the reduction of reaction rate after hydrogen loading should be negligible. We thus conclude that the reaction rate does not only depend on refractive index, as reported for Ge-doped fibers in [3], but that the proportionality factor between reaction rate and index depends on processing like hydrogen loading or UV irradiation.

Correlation of etch rate changes with core stress changes

As we show later in this chapter, an increase in etch rate with fiber drawing tension can be found for nitrogen-doped fibers drawn from the same preform. This result indicates an increase in etch rate with compressive core stress, as the fibers core stress was found to decrease linearly with drawing tension [1]. An increase of etch rate with compressive stress has also been reported by Agarwal and Tomozawa [12] for quenched bulk silica glass samples. However, for phosphorus-doped fibers, stress changes due to hydrogen loading were found to be smaller than ± 1 kg/mm^2 (error in stress measurements), and an increase in core stress was found after subsequent UV-irradiation [11]. If stress was the parameter dominating the reaction rate, we would thus expect no change in etch rate with hydrogen loading and a decreasing etch rate after UV-illumination. This is in contradiction with our experiments, so we conclude that stress does not influence the etch rate in our case.

Color centers and etch rate changes

Hydrogen loading as well as UV-irradiation alters the defect population of the glass. For phosphorus-doped glass samples, an increase in UV-absorption after hydrogen loading has been found for photon energies >5.5 eV [7, 13]. Furthermore, an increase in absorption due to POH formation after hydrogen loading has been reported for phosphorus-doped fibers at room temperature [14]. Hosono et al. explained the electron spin resonance (ESR) spectra of hydrogen loaded glass samples after irradiation with 6.4 eV photons by a superposition of the two spectra from paramagnetic PO$_2$ and PHO$_2$ centers [7]. The defects are created from hydrogen-induced precursors causing the absorption band for photon energies >5.5 eV. The PO$_2$ and PHO$_2$ centers involve a drastic coordination change around phosphorus, resulting in a densification of the glass [7].

We suggest that the reaction rate changes reported in this paper are governed by color center concentrations and their modification due to hydrogen loading and subsequent UV-illumination. The rate-limiting reaction for the dissolution of silica by hydrofluoric acid is the substitution of a surface SiOH group, which is bonded to three neighboring oxygen atoms, by a SiF group [15]. The same should hold true for dopants introduced in a silica glass matrix. A color center is a defect where at least one of the four bonds to neighboring oxygen atoms has been replaced by non-bridging oxygen, paired- or unpaired electrons, hydride-, or hydroxyl groups. It is directly evident that the exact nature of the defect can hamper or facilitate its substitution by fluorine considerably. Following this picture, hydrogen loading of the fiber results in a passivation of drawing induced defects, which hampers their substitution by fluorine. In contrast, the irradiation of the hydrogen loaded core glass results in the formation of PO_2 and PHO_2 centers [7], which are only linked to two bridging oxygen atoms. The drastic change in coordination number, accompanied by the formation of unpaired electrons, facilitates the adsorption of fluorine on the phosphorus atom, thus enhancing the etch rate.

The stress change of the core glass has not been found to depend significantly on pulse fluence, whereas the induced refractive index change more than doubles for a pulse fluence of 50 mJ/cm^2 as compared to a pulse fluence of 350 mJ/cm^2 [11]. It was concluded that the densification process does not explain the difference in refractive index for the two fluences. In this study, however, we could not either find a significant dependence of etch rate on pulse fluences. A possible explication for the index difference might be the creation of an additional color center for low fluences, which neither contributes to densification, nor affects the core etch rate.

In contrast to the results reported by Canning et al. [6], we did not observe any preferential etching at the core/cladding interface after UV-irradiation. The fiber used within this study has about the same phosphorus concentration as the fiber investigated in [6], but no details about irradiation conditions were given in the publication. The effect reported in [6] might thus only occur for a limited range of irradiation parameters.

6.1.5 Conclusions

The dependence of etch rate on dopant concentration for phosphorus-doped fibers has been investigated. Changes in etch rate due to hydrogen loading and subsequent UV-irradiation have been studied. The etch rate decreases by about 30 % after hydrogen loading and increases by about 30 % after UV-irradiation with respect to the pristine fiber. No etch rate dependence on pulse fluence is observed. We conclude that the reaction rate of the etching depends on processing like hydrogen loading and UV irradiation. AFM-measurements thus yield geometric information about the fiber index profile, but do not provide absolute information about the refractive index even for a single dopant. Changes in defect population are suggested to be the reason for the observed etching behavior. Etching of optical fibers can thus be used to get additional insight in the color center concentration and changes of the glass network.

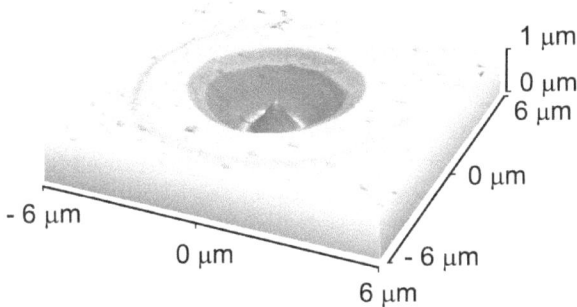

Figure 6.6: *AFM image of the nitrogen-doped fiber drawn with a tension of* 125 g *after* 5 *minutes of etching in* 40 % *HF.*

6.2 Nitrogen-doped fibers

6.2.1 Experiment

The fibers under investigation are four nitrogen-doped fibers drawn from the same preform at different drawing tensions. The fibers have been described in more detail in section 5.2.1 of the preceding chapter. There, a linear increase of core stress with both drawing tension and UV-irradiation has been determined.

All of the four fibers have been irradiated with an ArF excimer-laser using the same conditions as described in section 5.2.3 of the last chapter, i.e. with a pulse fluence of $(100 \pm 20)\,\mathrm{mJ/cm^2}$ and total doses of $1.2\,\mathrm{kJ/cm^2}$ and $2.4\,\mathrm{kJ/cm^2}$, respectively, resulting in 4 pristine and 8 irradiated fiber samples.

For AFM observations, the fibers have been cleaved and the resulting end-faces were etched in 40 % HF solution. Etch times from 1 to 5 minutes resulted in etch depth smaller than 1 µm, which is about the maximum dynamic range of the AFM. For every irradiation condition, at least three different etch times were used to determine the etch rate. The samples were glued on an aluminum mount and vertically aligned. The AFM (TopoMetrix Explorer) was operated in contact mode in air with a standard V-shaped silicon nitride cantilever.

6.2.2 Etch dynamics of the pristine fiber

Figure 6.6 shows the end-face topography of the nitrogen-doped fiber drawn with a tension of 125 g. The fiber was etched for 5 minutes. The nitrogen-doped core can clearly be identified due to its higher etching depth with respect to the surrounding silica tubes, resulting in a symmetric crater of 4 µm (FWHM) in diameter. The "burn off" region, caused by the collapsing process during preform fabrication, can clearly be observed as a dip in the center of the crater. In contrast to the results presented in [16], no changes of profile with drawing temperature indicating dopant diffusion could be observed. The silica supporting tube embedding the core is found to etch most slowly and is further taken as reference level.

Figure 6.7 illustrates the evolution of etch profiles over time for the fiber drawn with a tension of 125 g. A nonlinear dependence of etch depth on time is evident in the center of the core as well as for a radial position of $r = \pm 2$ µm. For the time range observed, etch rates can thus not only be determined by the dopant concentration as found for Ge-doped fibers in [17], but the end-face topography of the fiber must also influence the etch dynamics. Nevertheless, a linear dependence of etch depth on time is found for the etch maxima at $r = \pm 1$ µm and is illustrated in Fig. 6.8. The slope of the linear fit on the data determines the etch rate at the corresponding radial position. The etching behavior in the center of the core ($r = 0$ µm) is also illustrated in Fig. 6.8 for comparison. Up to an etch time of about 180 seconds, the relative etch depth equals zero, and shows an almost linear behavior for higher etch times.

The maximum etch rates of the four pristine fibers are plotted as a function of drawing tension in Fig. 6.9. Error bars indicate the significant confidence level. The etch rates of the pristine fibers increase with drawing tension, saturate, and then slightly decrease. The etch rate maximum is found for the drawing force of 195 g.

6.2.3 UV-induced etch rate changes

After UV-irradiation, the etch rate increases for all fibers under investigation (Fig. 6.9). The increase in etch rate does not depend significantly on pulse number for the pulse numbers observed within this experiment. It lies in between 6% for the fiber drawn at 195 g and 15% for the fiber drawn with 65 g. The irradiated fibers show qualitatively the same evolution of the end-face profile with etch time as illustrated for the pristine fiber in Fig. 6.7. Particularly, the topographies of the UV-treated samples do not exhibit an asymmetry due to UV side exposure as it was observed for hydrogen-loaded germanium-doped fibers by Inniss *et al.* in [18]. For all fiber samples, the etch

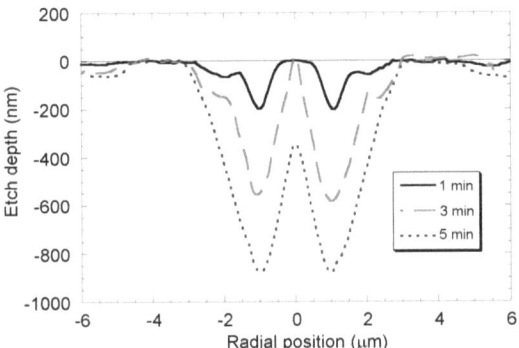

Figure 6.7: Etch dynamics of the nitrogen-doped fiber drawn with a tension of 125 g. Etch times were 1, 3 and 5 minutes. A linear dependence of etch depth on time is found for the etch maxima at a radial position of $r = \pm 1\,\mu m$.

depth maxima at a radial position of $r = \pm 1$ μm are found to depend linearly on time, so that the corresponding etch rates can directly be determined for further comparison.

6.2.4 Discussion

Etch dynamics

The results for the AFM observations reported above disagree in two aspects with results published so far. Firstly, the etch profiles shown in Fig. 6.7 do not exhibit a linear dependence on etch time, as has been reported for germanium-doped fibers by Huntington *et al.* in [17]. The nonlinear dependence of etch depth on etch time can be explained by the influence of surface curvature on the etch rate. Sharp edges with convex curvature etch faster than flat surfaces due to additional lateral attack of the material. The disappearance of the peaks at a radial position of $r = \pm 2\,\mu m$ in Fig. 6.7 can thus be explained as well as the threshold in etch rate in the center of the core. There, the etching starts only after the originally flat surface has become pointed (Fig. 6.8). The profile for the lowest etch time (1 min) in Fig. 6.7 should therefore give the most detailed information about the radial distribution of the glass properties defining the etch rate, as the influence of topography on the etch rate is still small for short times.

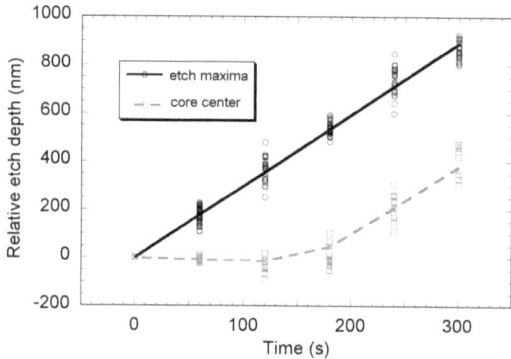

Figure 6.8: *Relative etch depth of the nitrogen-doped fiber as a function of time for the etch maxima at $r = \pm 1\,\mu m$ and for the center of the core. The etch maxima depend linearly on time. Drawing tension was 125 g.*

Etch rate changes with drawing tension

As depicted in Fig. 6.9, the etch rate varies by more than 25 % for the four nitrogen-doped fibers drawn with different drawing tensions. For all fibers, however, the dopant concentration is the same as they have been drawn from the same preform. Furthermore, no change in index has been found within the measurement accuracy of ±5 %. The etch rate is thus not only determined by the dopant concentration or the refractive index. It must rather be influenced significantly by at least one more parameter that is modified in the fiber drawing process.

During the fiber fabrication process, the drawing tension has been altered due to a change in temperature, whereas all other parameters have been kept constant. Fiber properties depending on drawing temperature and tension can be both structural and chemical in nature, where we relate structural properties to the ensemble of atoms forming the glass matrix, whereas chemical properties are attributed to certain localized deviations from the ideal configuration, so-called color-centers.

The residual stress frozen into a fiber during the drawing process is a structural property in this sense. A linear dependence of axial stress on drawing tension has been demonstrated for the fibers under investigation in the last chapter (Fig. 5.14). The higher the tension, the higher the compressive core stress. However, the dependence of etch rate on drawing tension is found to be nonlinear (Fig. 6.9). We thus conclude that axial stress does not dominate the observed etch rate changes.

Other structural parameters modified during fiber fabrication are fictive

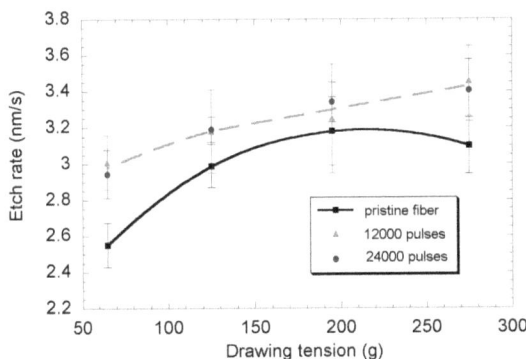

Figure 6.9: *Etch rate dependence of the pristine and UV-exposed nitrogen-doped fiber as a function of drawing tension. UV-illumination enhances the etch rate for all samples under investigation.*

temperature and Rayleigh scattering. A linear relation between these two parameters has been reported in [19]. The higher the fictive temperature, the higher is the Rayleigh scattering coefficient of the sample, independently of the samples thermal history. Furthermore, a linear increase of the Rayleigh scattering coefficient with drawing temperature has been established in [20]. As drawing tension decreases with increasing drawing temperature, Rayleigh scattering as well as fictive temperature decrease with drawing tension. In [12], an increase in etch rate has been observed for increasing fictive temperature. We thus conclude that, in our case, etch rate should decrease with increasing drawing tension, i.e. decreasing drawing temperature. However, as can be seen in Fig. 6.9, the opposite is true. Thus, fictive temperature and Rayleigh scattering do not correlate with the etch characteristics of the fibers under investigation.

During fiber fabrication, a variety of different color-centers are introduced in the fiber. The structure of the defects as well as their dependence on drawing conditions have been studied extensively for pure silica fibers [21–23]. In general, two different origins of defect generation have been established: thermal activation [21, 23] and shear stress [22]. The thermally activated defects grow in population with increasing drawing temperature, whereas stress-induced defects grow over drawing tension. As drawing tension increases with decreasing temperature, the total number of defects thus depends on two opponent processes. If drawing-induced color-center modifications were the origin of the observed etch rate changes, this might explain that the etch rate exhibits a maximum for a drawing tension of about 195 g (Fig. 6.9).

However, as no detailed information about the structure and generation of defects for nitrogen-doped fibers has been reported so far, correlation with experimentally observed color-center concentrations is not possible at this moment.

UV-induced etch rate changes

The changes in etch rate due to UV-illumination for irradiations with 12000 and 24000 pulses are illustrated in Fig. 6.9. After UV-irradiation, the etch rate has been increased by about 10 % in average in comparison to the pristine fibers. Etch rate changes between 12000 and 24000 pulses cannot be distinguished within the error of the measurement.

As refractive index changes for comparable irradiation conditions are only about 10^{-4} [24], the observed etch rate changes do not correlate with refractive index changes. In addition, the results confirm the negligible influence of stress changes on etch rate. An increase of etch rate with compressive stress has been reported for quenched silica glass samples by Agarwal et al. in [12]. UV-irradiation, however, reduces the compressive core stress, as we demonstrated in the last chapter. If stress was influencing the etch rate, it should thus be decreased and not increased by UV-irradiation.

The saturation of etch rate over pulse number depicted in Fig. 6.9 might rather be another indication that the etch rate correlates with the concentration of color centers. For nitrogen-doped fibers, absorption band changes induced by 193 nm irradiation have been found and attributed to a modification of nitrogen-associated color centers [25]. Lee at al. reported a modification of processing-induced defects by UV-irradiation in pure silica fibers [26]. The changes in defect population were found to occur predominantly during the first few hundreds of pulses and to slow down for higher pulse numbers. Qualitatively, we observe the same phenomenon in our case. However, as little has been published about the nature of defects in nitrogen-doped glasses, we cannot assign the etch rate to one specific defect in the fiber core. The changes in etch rate might even be caused by different color centers, whose concentrations have been redistributed due to interaction with UV-irradiation. A variety of defects are characterized by unpaired electrons, which could facilitate the adsorption of hydrofluoric acid on the glass and thus speed up the etching process. As for the phosphorus-fiber observed in the previous section, we thus conclude that the etch rate changes are due to a UV-triggered redistribution of drawing-induced defects.

The comparison of etch rate and stress changes over pulse number does also give additional information about the different contributions to photosensitivity in nitrogen-doped fibers. As the etch rate saturates over pulse number, we conclude that changes in color centers occur predominately at the beginning of the irradiation. In contrast, no saturation can be observed for stress changes during irradiation (cf. Fig 5.20). The densification-induced contribution to the overall index change might thus dominate the index change for higher total fluences.

6.2.5 Conclusions

Changes in etch rate due to UV-irradiation have been determined for four nitrogen-doped fibers drawn from the same preform with different drawing tensions using an AFM. A nonlinear dependence of etch rate on drawing tensions was found. UV-illumination enhances the etch rate by $6-15\,\%$ for all fibers observed. No significant influence of pulse number on etch rate could be established for 12000 and 24000 pulses, respectively. We suggest that the etch rate is influenced by color centers introduced in the fiber during the fabrication process and modified by UV-illumination. The main contribution of color centers to photosensitivity thus occurs at the beginning of the irradiation. The increase in refractive index can thus be explained by the superposition of two effects, where the first one, redistribution of color centers, occurs on a shorter timescale as the second one, the densification of the doped core region.

6.3 Summary and Discussion

The local etch rate of cleaved fiber end-faces was determined by measuring topography changes due to etching as a function of time with an AFM. The etch speed was found to depend strongly on the dopant, so the concentration of hydrofluoric acid was adapted to get reasonable etching times. For the fiber doped with 1 at% nitrogen, a HF-concentration of $40\,\%$ and an etch time of 5 minutes was necessary to get an etch depth of about 800 nm, whereas about the same depth was achieved with a $5\,\%$ solution and only 3 minutes of etch time for the fiber doped with phosphorus. The nitrogen-doped fiber did not show a linear dependence of etch depth on time at certain radial positions, which might be explained by the influence of the surface curvature of the sample on the etch speed. As comparable effects have not been found for germanium- or phosphorus fiber, they might only appear when the fiber core is highly resistant to the chemical attack, so that high acid concentrations must be used.

For the phosphorus-doped fiber, etch rate reductions of about $30\,\%$ due to hydrogen-loading, and an increase of about the same amount due to UV-irradiation have been found. We explained the hydrogen-induced reduction of etch rate with a corresponding passivation of color-centers by hydrogen.

UV-irradiation, on the other hand, generates new color-centers and thus increases the chemical activity.

For nitrogen-doped fibers, a nonlinear dependence of etch rate on drawing tension was observed, which can also be attributed to a dependence of color-center concentrations on the fiber drawing conditions. The etch rate was increased by 10 % in average by UV-irradiation, which again was explained by a corresponding redistribution of the color center population.

The etch rate changes for the phosphorus-doped fibers are significantly higher than those for the nitrogen- or for germanium-doped [18] fibers. Complementary, we approximated the densification-induced index change to be smaller in phosphorus- than in nitrogen- or germanium-doped fibers (cf. table 5.1 in chapter 5). The amount of etch rate changes due to UV-irradiation might thus provide additional information about the different contributions to the overall index change.

Bibliography

[1] F. Dürr, G. Jänchen, H. G. Limberger, and S. L. Semjonov. Atomic Force Microscopy Study of UV-irradiated fibers drawn at different drawing tensions. In *Summer School on Photosensitivity in Optical Waveguides and Glasses (POWAG 2002)*, 2002. paper MoA8.

[2] F. Dürr, H. G. Limberger, R. P. Salathé, S. A. Vasiliev, O. I. Medvedkov, A. S. Bozhkov, and E. M. Dianov. Annealing-induced stress changes in UV-irradiated germanium-doped fibers. In *BGPP 2005 Technical Digest*. Optical Society of America, 2005. accepted as poster presentation.

[3] P. Pace, S. T. Huntington, K. Lyytikainen, A. Roberts, and J. D. Love. Refractive index profiles of Ge-doped optical fibers with nanometer spatial resolution using atomic force microscopy. *Optics Express*, 12(7):1452–1457, 2004.

[4] Q. Zhong and D. Inniss. Characterization of the lightguiding structure of optical fibers by atomic force microscopy. *Journal of Lightwave Technology*, 12(9):1517–1523, 1994.

[5] M. M. Bubnov, E. M. Dianov, O. N. Egorova, S. L. Semjonov, A. N. Guryanov, V. F. Khopin, and E. M. DeLiso. Fabrication and investigation of single-mode highly phosphorus-doped fibers for Raman lasers. *Proceedings of the SPIE*, 4083:12–22, 2000.

[6] J. Canning, K. Sommer, M. Englund, and S. Huntington. Direct evidence of two types of UV-induced glass changes in silicate based optical fibers. *Advanced Materials*, 13(12-13):970–973, 2001.

[7] H. Hosono, K. Kajihara, and M. Hirano. Photochemistry in phosphorous-doped silica glass by ArF excimer laser irradiation: crucial effect of H_2 loading. *Journal of Applied Physics*, 91(7):4121–4124, 2002.

[8] W. Hermann, M. Hutjens, and D. U. Wiechert. Stress in optical waveguides. III. Stress induced index change. *Applied Optics*, 28(11):1980–1983, 1989.

[9] N. H. Ky, H. G. Limberger, R. P. Salathé, F. Cochet, and L. Dong. Effects of drawing tension on the photosensitivity of Sn-Ge- and B-Ge-codoped core fibers. *Optics Letters*, 23(17):1402–1404, 1998.

[10] N. H. Ky, H. G. Limberger, R. P. Salathé, F. Cochet, and L. Dong. Hydrogen-induced reduction of axial stress in optical fiber cores. *Applied Physics Letters*, 74(4):516–518, 1999.

[11] F. Dürr, H. G. Limberger, R. P. Salathé, F. Cochet, A. A. Rybaltovsky, Y. V. Larionov, S. L. Semjonov, and E. M. Dianov. UV-induced stress changes in phosphorus-doped fibers drawn at different drawing tensions. In *BGPP 2003 Technical Digest*. Optical Society of America, 2003. paper BThA2.

[12] A. Agarwal and M. Tomozawa. Correlation of silica glass properties with the infrared spectra. *Journal of Non-Crystalline Solids*, 209(1–2):166–174, 1997.

[13] B. Malo, J. Albert, F. Bilodeau, T. Kitagawa, D. C. Johnson, K. O. Hill, K. Hattori, Y. Hibino, and S. Gujrathi. Photosensitivity in phosphorus-doped silica glass and optical waveguides. *Applied Physics Letters*, 65(4):394–396, 1994.

[14] K. W. Plessner and S. J. Stannard-Powell. Attenuation/time telation for OH formation in optical fibers exposed to H_2. *Electronics Letters*, 20(6):250–252, 1984.

[15] D. M. Knotter. Etching mechanism of vitreous silicon dioxide in HF-based solutions. *Journal of the American Chemical Society*, 122:4345–4351, 2000.

[16] K. Lyytikäinen, S. T. Huntington, A. L. G. Carter, P. McNamara, J. W. Fleming, J. Abramczyk, I. Kaplin, and G. Schötz. Dopant diffusion during optical fibre drawing. *Optics Express*, 12(6):972–977, 2000.

[17] S. T. Huntington, P. Mulvaney, A. Roberts, K. A. Nugent, and M. Bazylenko. Atomic force microscopy for the determination of refractive index profiles of optical fibers and waveguides: a quantitative study. *Journal of Applied Physics*, 82(6):2730–2734, 1997.

[18] D. Inniss, Q. Zhong, A. M. Vengsarkar, W. A. Reed, S. G. Kosinski, and P. J. Lemaire. Atomic force microscopy study of UV-induced anisotropy in hydrogen-loaded germanosilicate fibers. *Applied Physics Letters*, 65(12):1528–1530, 1994.

[19] S. Sakaguchi, S. Todoroki, and T. Murata. Rayleigh scattering in silica glass with heat treatment. *Journal of Non-Crystalline Solids*, 220(2–3):178–186, 1997.

[20] K. Tsujikawa, K. Tajima, and M. Ohashi. Rayleigh scattering reduction method for silica-based optical fiber. *Journal of Lightwave Technology*, 18(1):1528–1532, 2000.

[21] H. Hanafusa, Y. Hibino, and F. Yamamoto. Formation mechanism of drawing-induced E' centers in silica optical fibers. *Journal of Applied Physics*, 58(3):1356–1361, 1985.

[22] Y. Hibino and H. Hanafusa. Defect structure and formation mechanism of drawing-induced absorption at 630 nm in silica optical fibers. *Journal of Applied Physics*, 60(5):1797–1801, 1986.

[23] Y. Hibino and H. Hanafusa. Formation mechanism of defect centers in GeO_2-doped silica glass. *Journal of Non-Crystalline Solids*, 95–96(1):343–350, 1987.

[24] E. M. Dianov, K. M. Golant, R. R. Khrapko, A. S. Kurkov, B. Leconte, M. Douay, P. Bernage, and P. Niay. Grating formation in a germanium free silicon oxynitride fibre. *Electronics Letters*, 33(3):236–238, 1997.

[25] K. M. Golant, E. M. Dianov, R. R. Khrapko, and A. L. Tomashuk. Nitrogen-doped silica fibers and fiber-based optoelectronic components. *Proceedings of the SPIE*, 4083:2–11, 2000.

[26] J. W. Lee, G. H. Sigel Jr., and J. Li. Processing-induced defects in optical waveguide materials. *Journal of Non-Crystalline Solids*, 57:57–65, 1998.

Chapter 7
Conclusions

7.1 Summary

This thesis has sought to provide information about laser-induced refractive index changes in optical fibers. Particularly, the contribution of compaction to photosensitivity has been studied in more detail. As core glass compaction is linearly related to an increase in the axial core stress, a setup capable of measuring stress changes in optical fibers with high accuracy and spatial resolution has been developed. The setup basically determines the integrated transverse birefringence of the fiber under investigation. Traditionally, the acquired birefringence profiles have been interpreted only to depend on stress. In this work, however, we also consider the effect of inelastic strains introduced into the fiber during its fabrication process. The birefringence can thus be separated in a constant, inelastic strain-induced part, and a radially changing stress-induced part. The setup is also capable to determine asymmetric stress profiles tomographically.

The setup was used to enlarge the insight in laser-induced compaction for germanium-doped fibers. Standard SMF-28® fibers have been irradiated with infrared femtosecond-laser pulses. Resulting asymmetric stress changes in the core were recorded tomographically. The relation between compaction-induced and overall index change was found to be comparable to fibers irradiated with pulsed UV-laser sources. Furthermore, highly germanium-doped fibers were exposed to continuous wave UV-irradiation from a frequency-doubled Argon-Ion laser. Again, the axial stress and thus the core densification were found to depend linearly on the related index change. As for the other irradiation sources and germanium-doped fibers, the contribution of overall to densification-induced index change was found to be about 60 %. In addition, the annealing behavior of the irradiated highly germanium-doped fibers was studied.

Drawing-induced stress and inelastic strain changes have been studied for phosphorus- and nitrogen-doped fibers drawn at different drawing tensions. Both residual stress and inelastic strain anisotropy were found to increase linearly with drawing tension for both fibers. Furthermore, stress changes due to irradiation with pulsed excimer laser irradiation at 193 nm have been observed and correlated with the overall index change. The densification-induced contribution to the overall index change was found to be comparable to germanium-doped fibers for the nitrogen-doped samples, whereas it proves to be one order of magnitude smaller for the hydrogenated phosphorus-doped fibers.

The etching behavior of phosphorus- and nitrogen-doped fibers have been studied using an atomic force microscope. For phosphorus-doped fibers, strong etch rate reductions of about 30 % due to hydrogen-loading and an increase in etch rate of about the same amount due to UV-irradiation has been found. For the nitrogen-doped fibers, nonlinear etch rate changes with drawing tension and an increase of about 10 % due to UV-irradiation could be observed. The changes in etch rate have been attributed to a redistribution of color centers by UV-irradiation.

7.2 Future work

Optical fiber photosensitivity remains a field of ongoing research. Due to the multitude of parameters influencing the amount of index change with irradiation, lots of further experiments are thinkable. For instance, the role of hydrogen-loading on compaction remains an important open question for germanium-doped fibers. Another interesting task is the correlation of compaction with color-centers. Thus, it could be investigated whether the two contributions are uncorrelated, or whether changes in certain color centers might be necessary to trigger the densification process.

The stress measurement setup realized within this work is limited so far to the measurement of stress profiles in optical fibers. A promising task might be the extension of the method to planar waveguide structures, as they have rising significance both in the domain of sensing and telecommunications.

Die VDM Verlagsservicegesellschaft sucht für wissenschaftliche Verlage abgeschlossene und herausragende

Dissertationen, Habilitationen, Diplomarbeiten, Master Theses, Magisterarbeiten usw.

für die kostenlose Publikation als Fachbuch.

Sie verfügen über eine Arbeit, die hohen inhaltlichen und formalen Ansprüchen genügt, und haben Interesse an einer honorarvergüteten Publikation?

Dann senden Sie bitte erste Informationen über sich und Ihre Arbeit per Email an *info@vdm-vsg.de*.

Sie erhalten kurzfristig unser Feedback!

VDM Verlagsservicegesellschaft mbH
Dudweiler Landstr. 99
D - 66123 Saarbrücken
www.vdm-vsg.de

Telefon +49 681 3720 174
Fax   +49 681 3720 1749

Die VDM Verlagsservicegesellschaft mbH vertritt

Printed by Books on Demand GmbH, Norderstedt / Germany